INSECTS & FLOWERS

THE ART OF MARIA SIBYLLA MERIAN

David Brafman | Stephanie Schrader

J. PAUL GETTY MUSEUM
LOS ANGELES

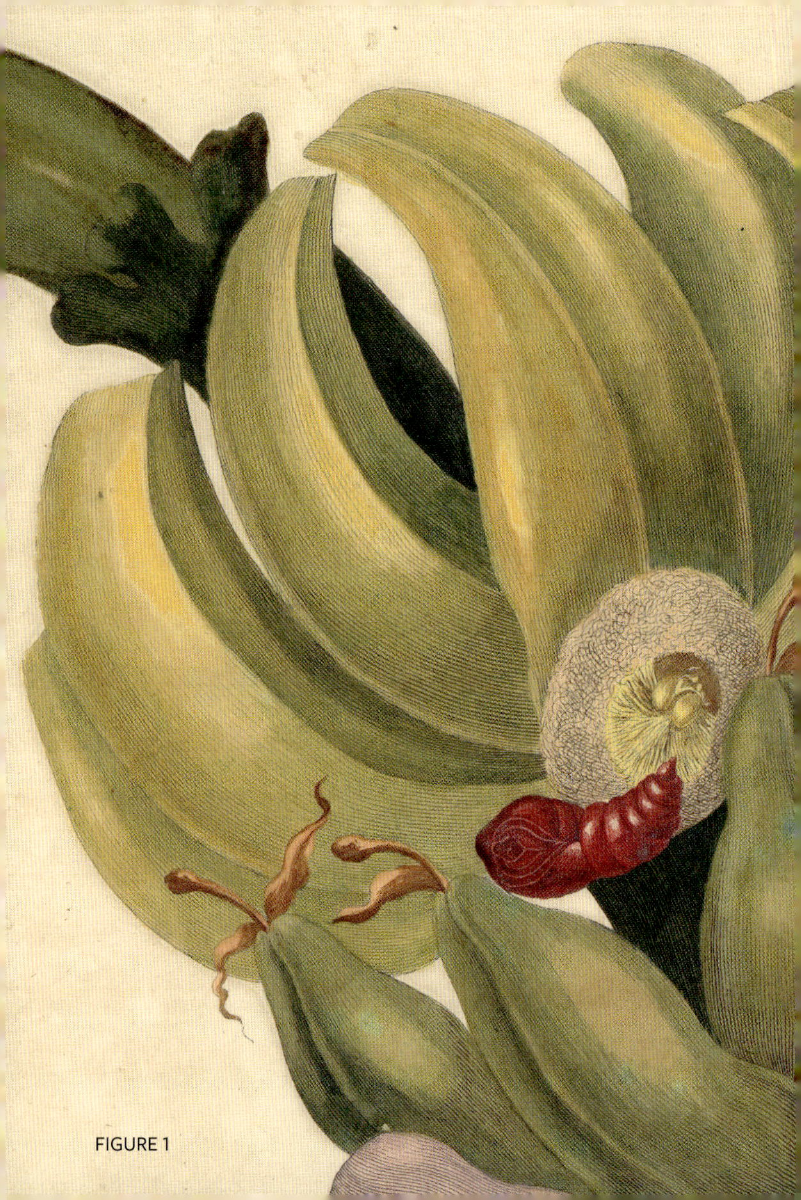

FIGURE 1

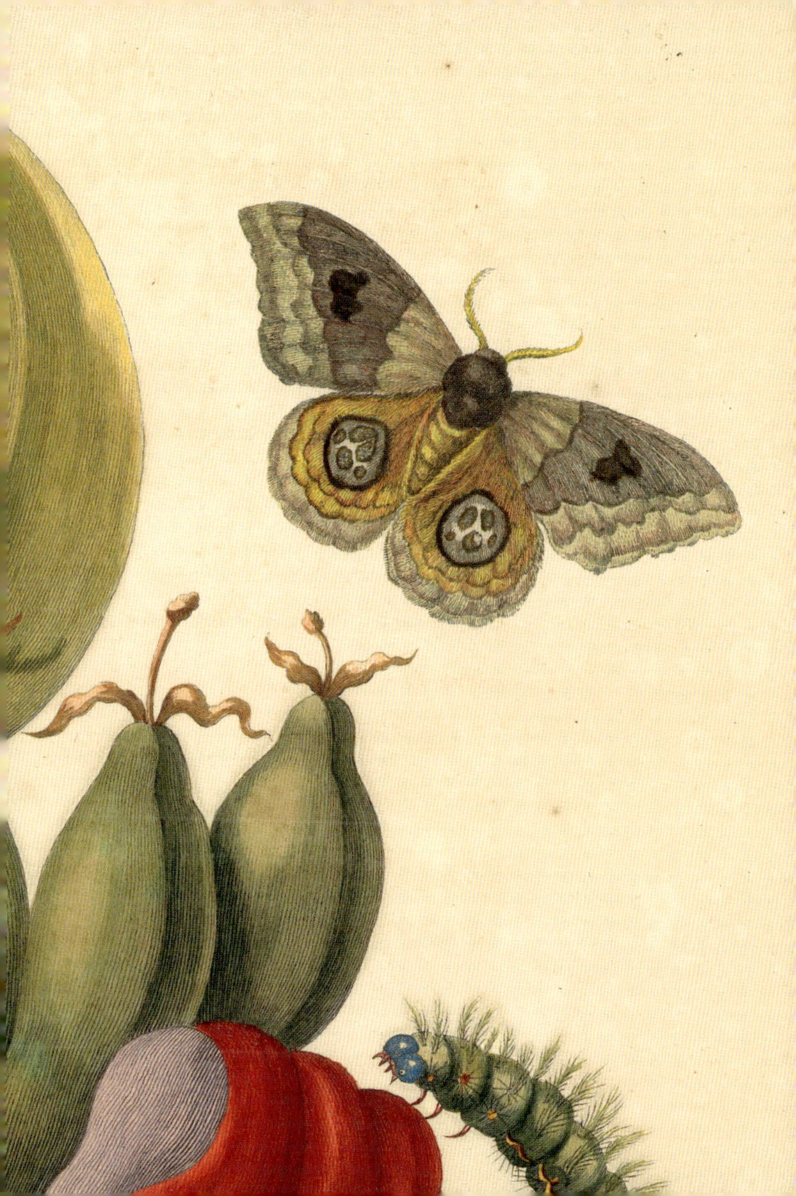

Sixth printing

PUBLISHED BY THE J. PAUL GETTY MUSEUM, LOS ANGELES
Getty Publications
1200 Getty Center Drive, Suite 500
Los Angeles, California 90049-1682
getty.edu/publications

Jesse Zwack, EDITOR
Stuart Smith, DESIGNER
Suzanne Watson, PRODUCTION COORDINATOR
Jobe Benjamin, John Kiffe, PHOTOGRAPHERS

Distributed in the United States and Canada by the University of Chicago Press
Distributed outside the United States and Canada by Yale University Press, London

Printed in China

Library of Congress Control Number: 2007940526

ISBN 978-0-89236-929-4

NOTE TO THE READER

Most of the illustrations in this book are full-scale details reproduced from hand-colored transfer prints in the second edition of Maria Sibylla Merian's *Metamorphosis of the Insects of Suriname* held by the Getty Research Institute (ID # 89-B10750).

This edition was published in Dutch as *Over de voortteeling en wonderbaerlyke veranderingen der Surinaemsche insecten* (Amsterdam, Joannes Oosterwyk, 1719). The engraved frontispiece and the seventy-two plates in the Getty Research Institute copy were magnificently hand colored near the time of its publication. This copy is bound in richly gold-tooled eighteenth-century mottled calf together with hand-colored 1730 editions of Merian's *Caterpillar Book* and *New Book of Flowers.*

The book was acquired in 1989 under the auspices of Anne-Mieke Halbrook, former Librarian, and Marcia Reed, former Curator of Rare Books, of the Getty Research Institute.

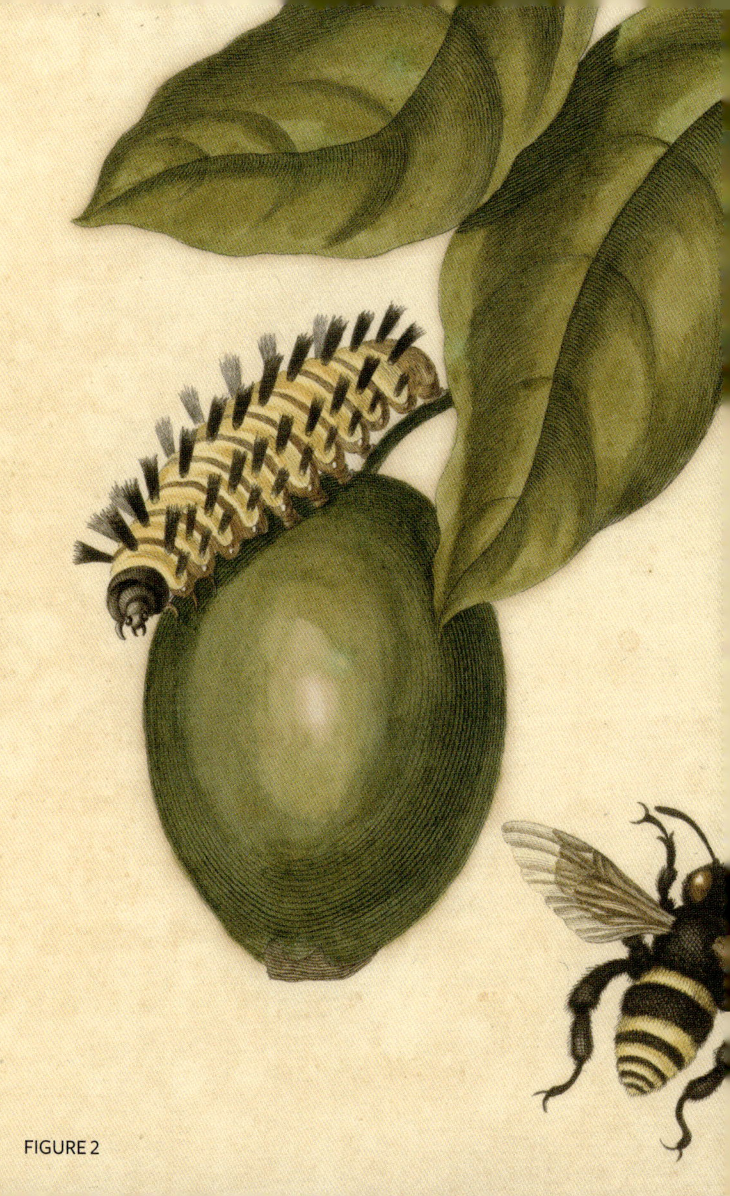

FIGURE 2

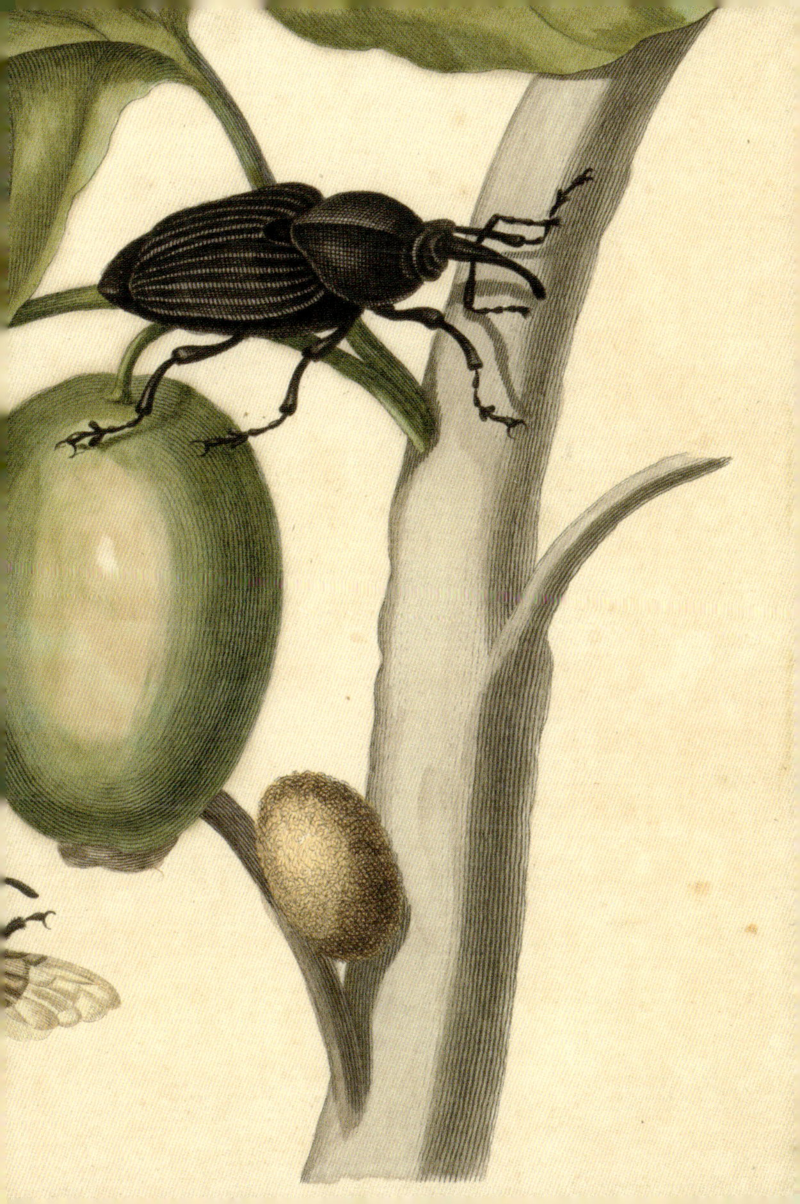

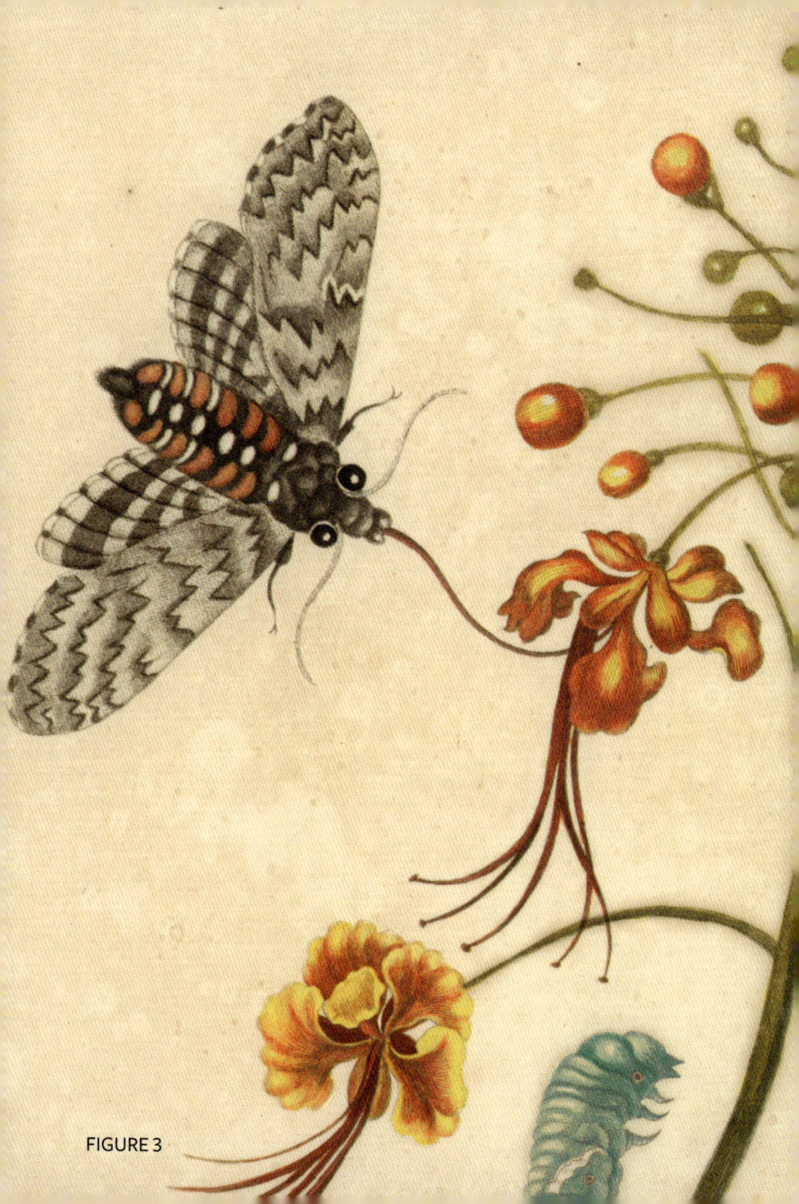

FIGURE 3

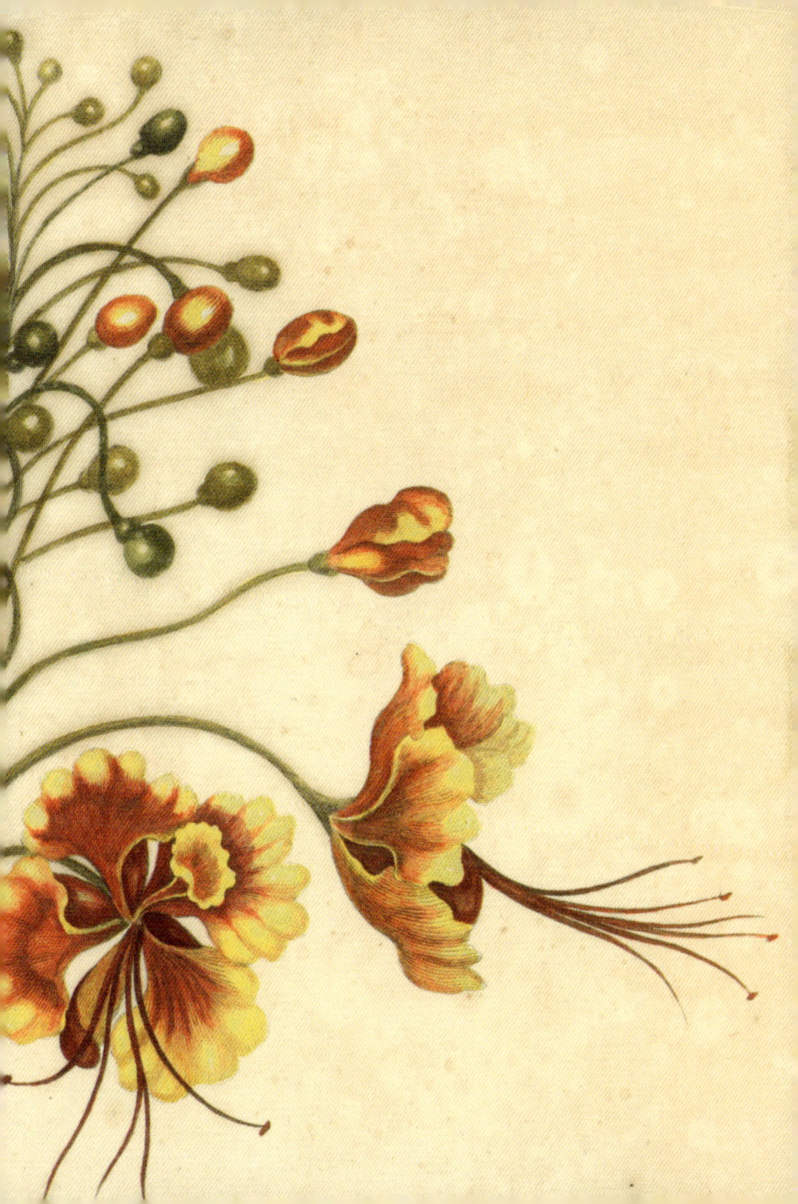

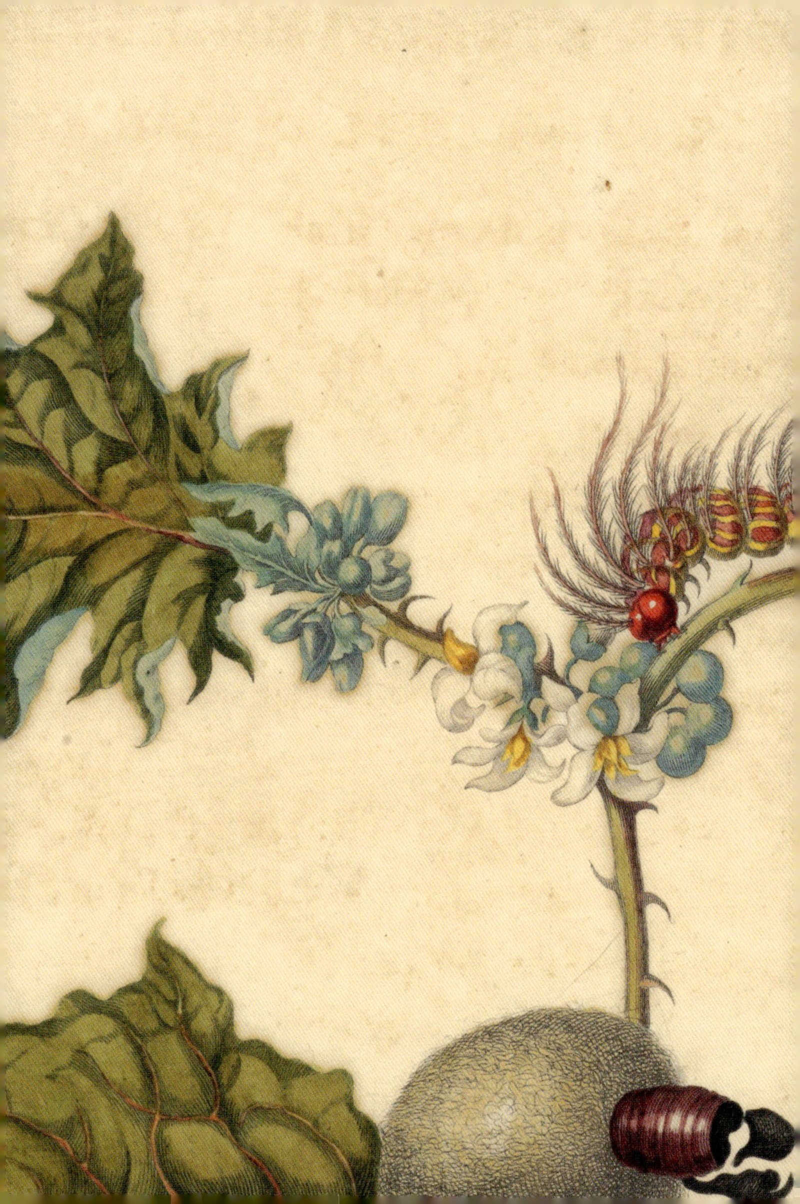

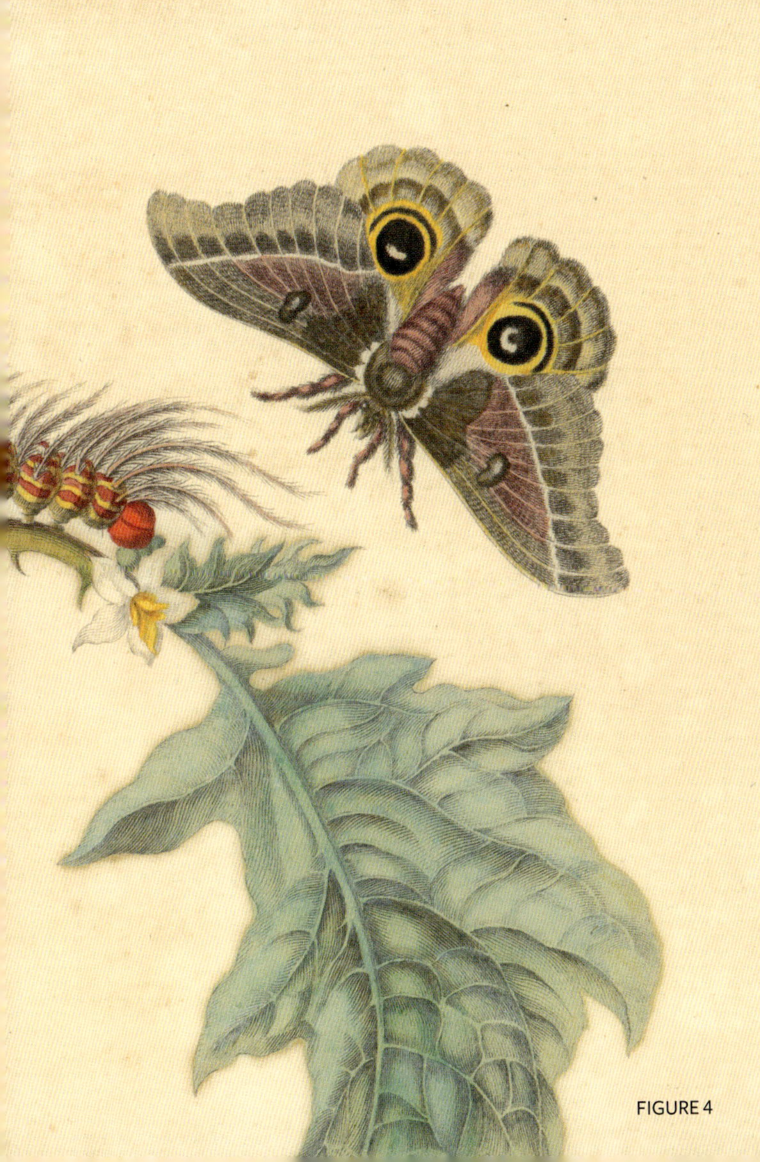

FIGURE 4

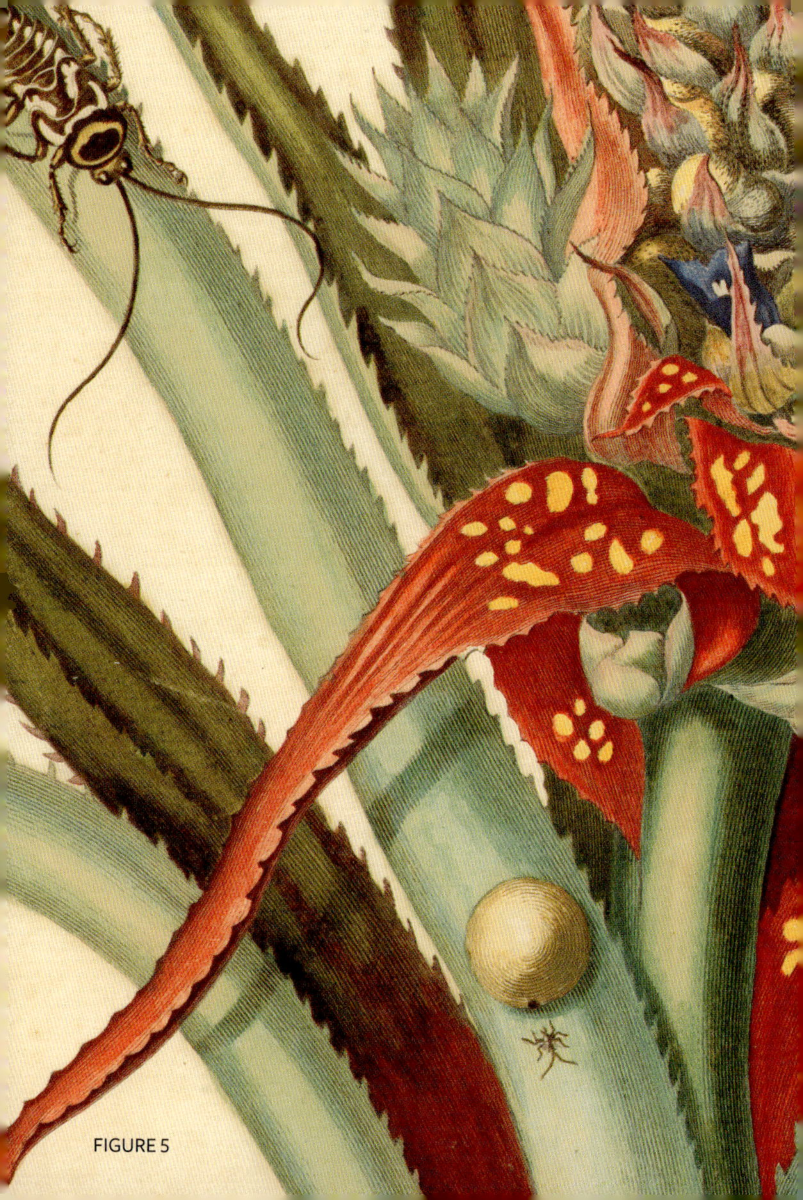

FIGURE 5

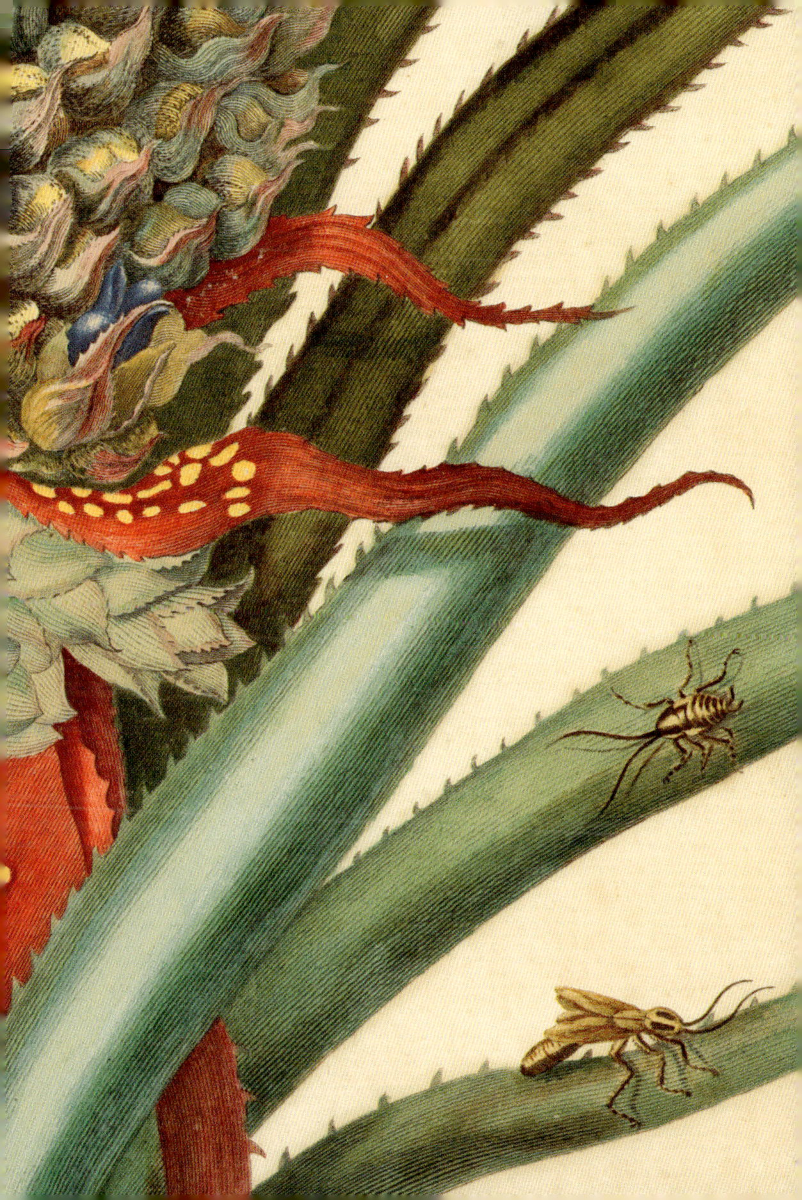

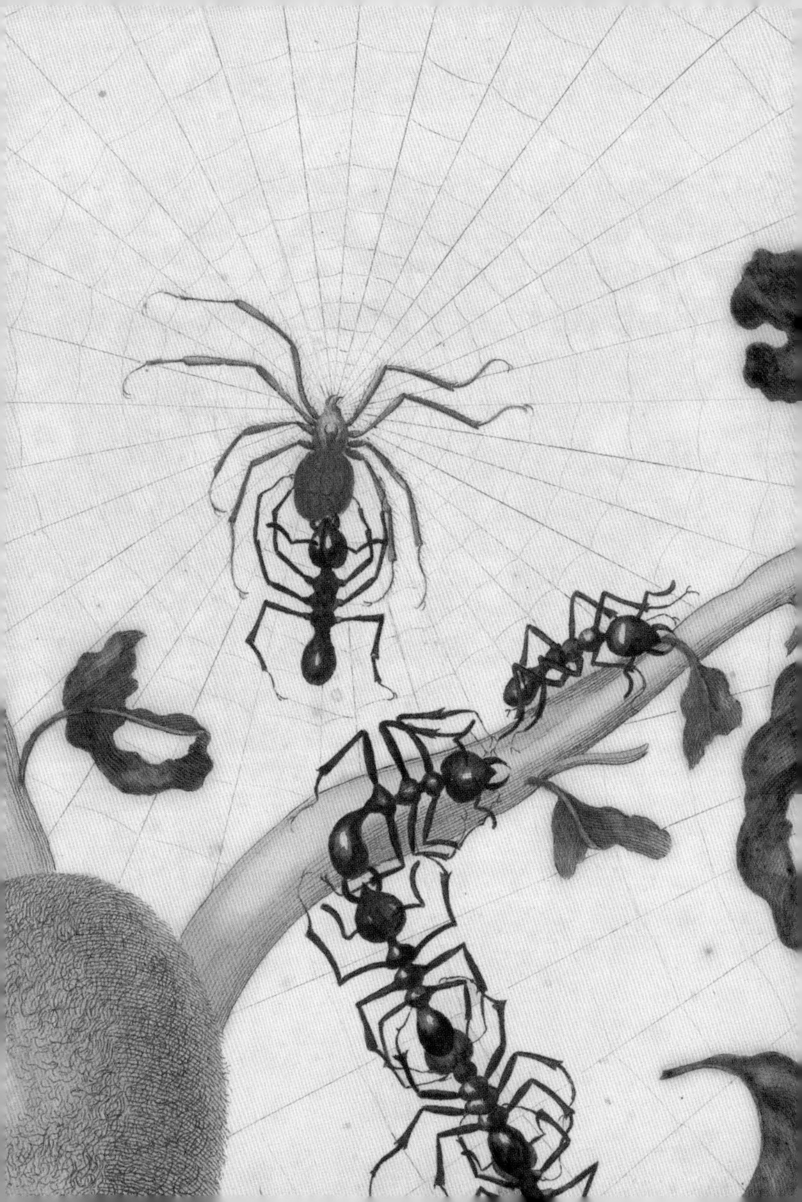

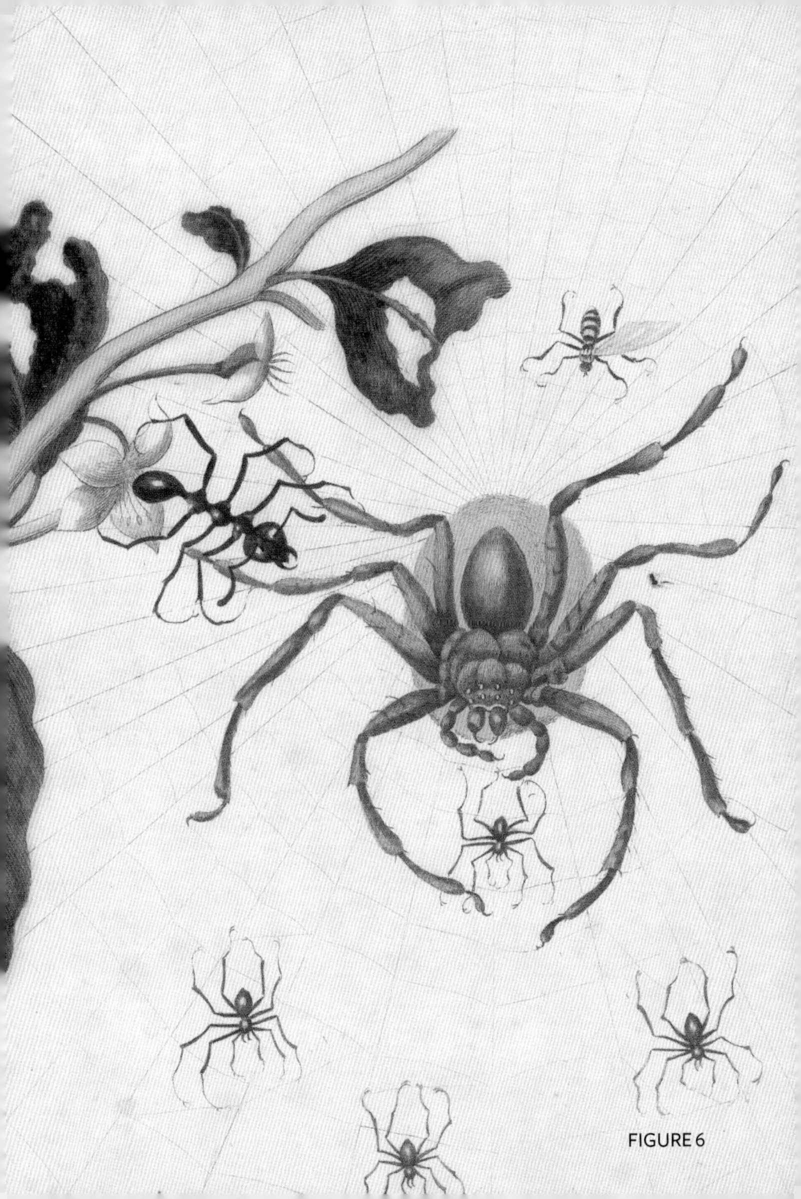

FIGURE 6

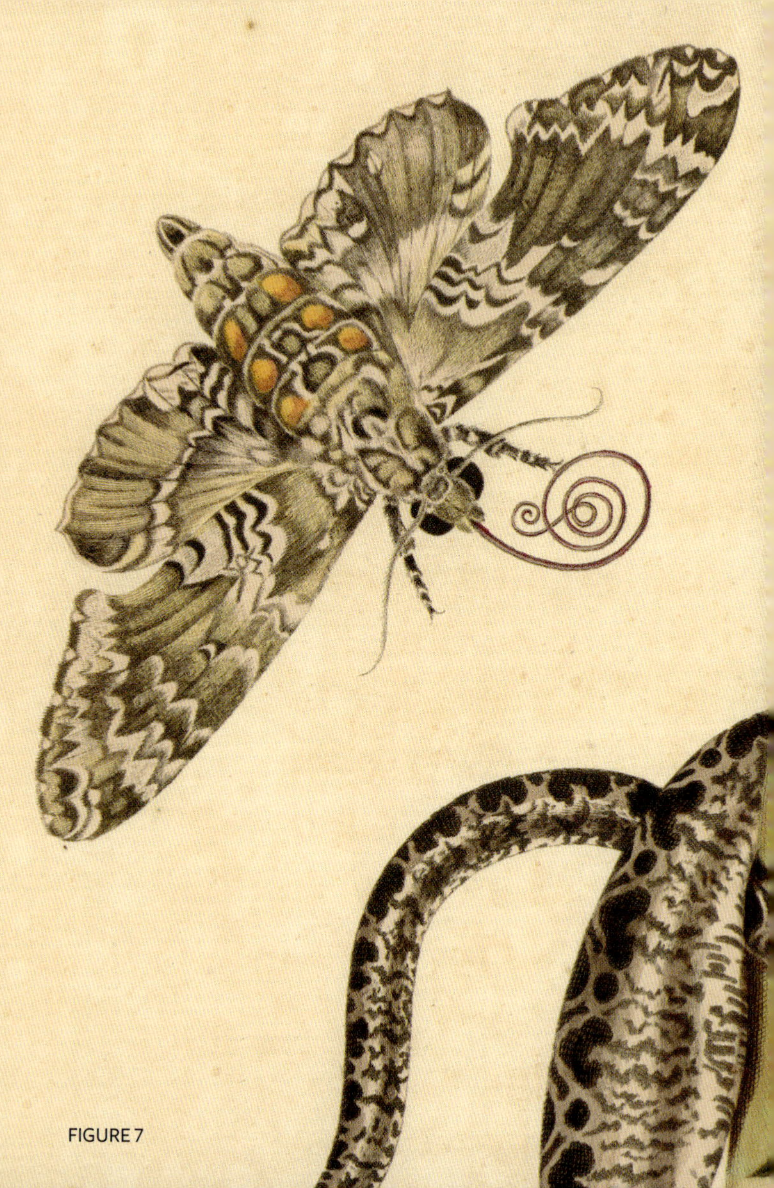

FIGURE 7

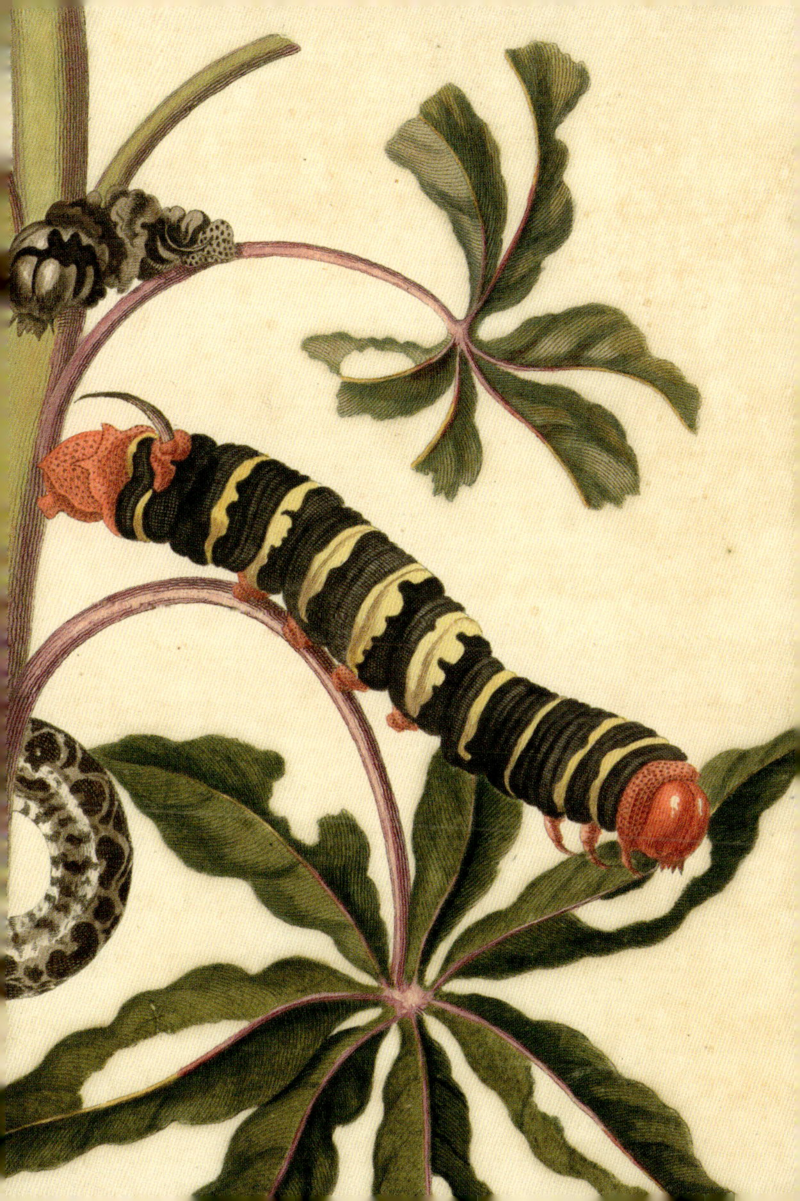

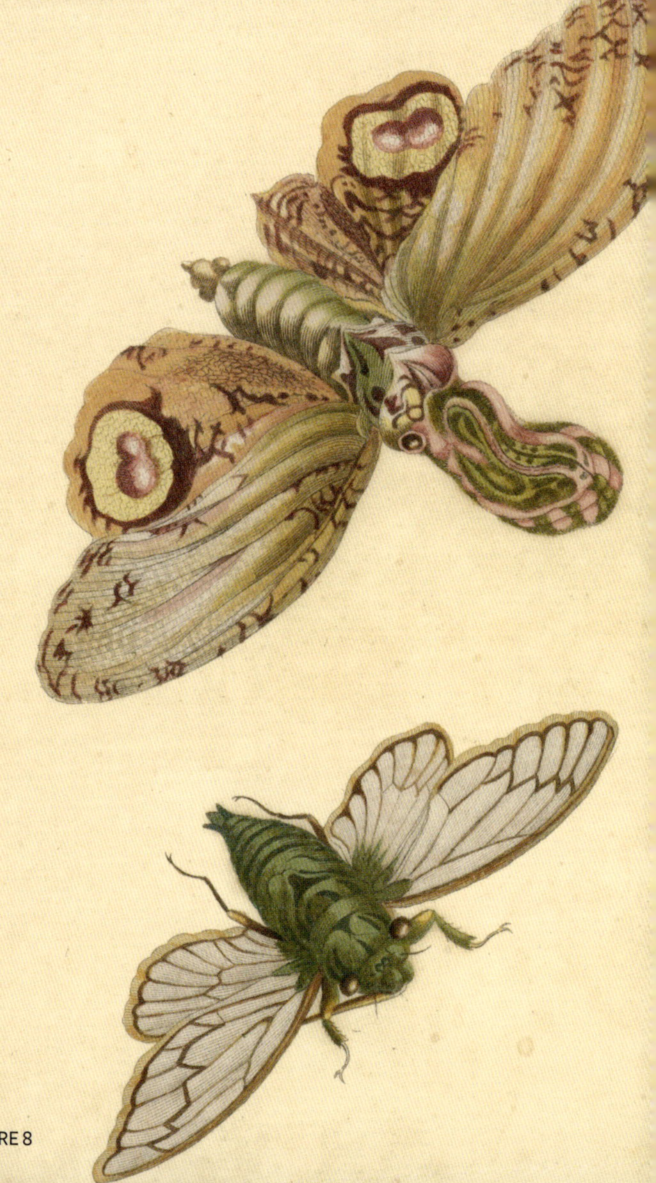

FIGURE 8

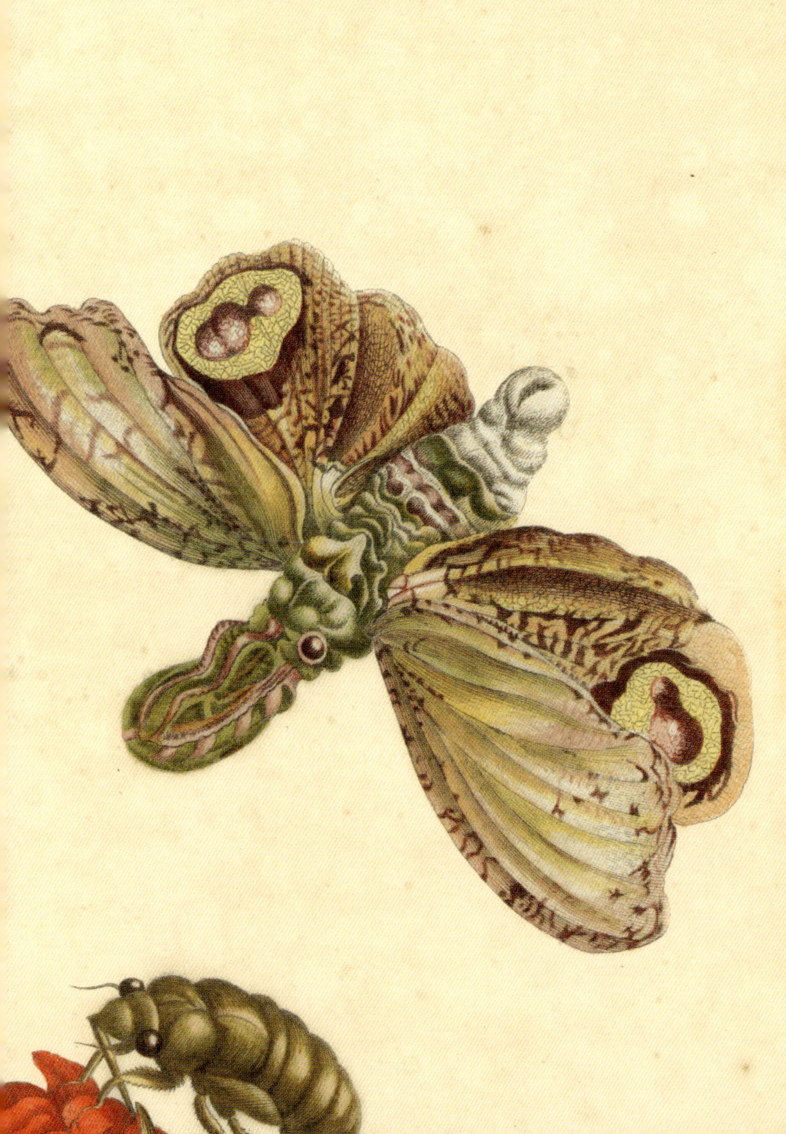

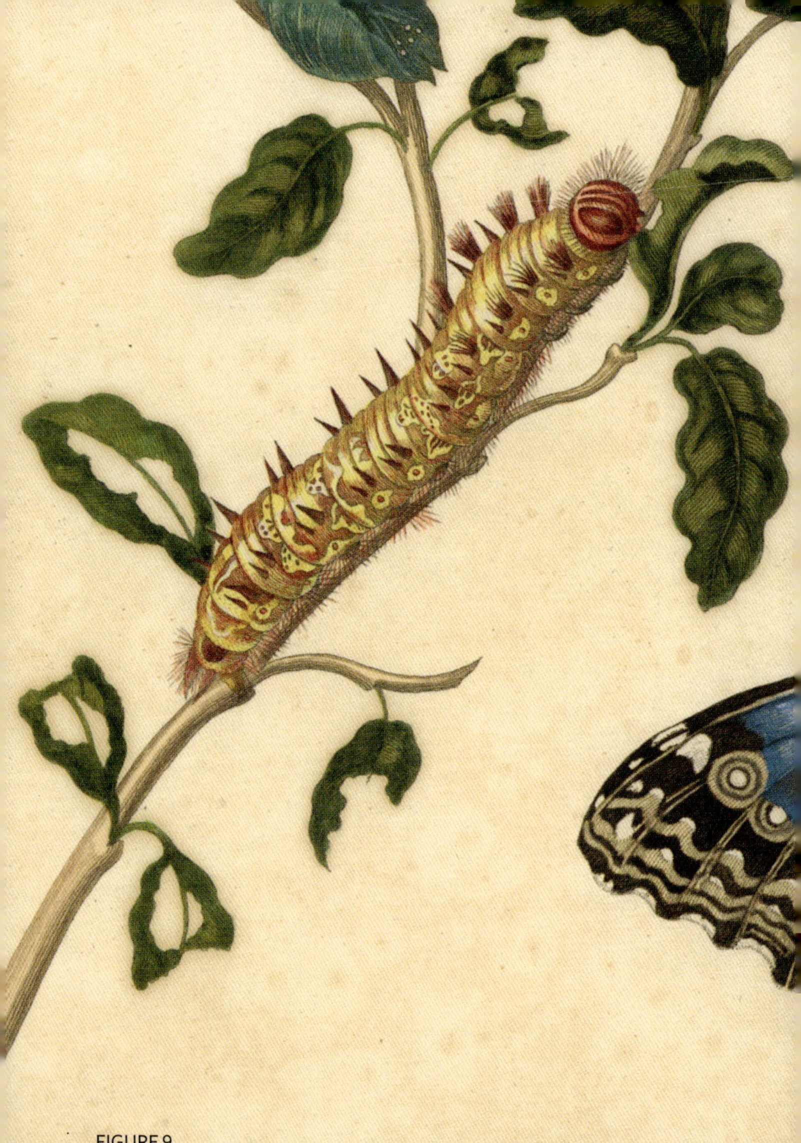

FIGURE 9

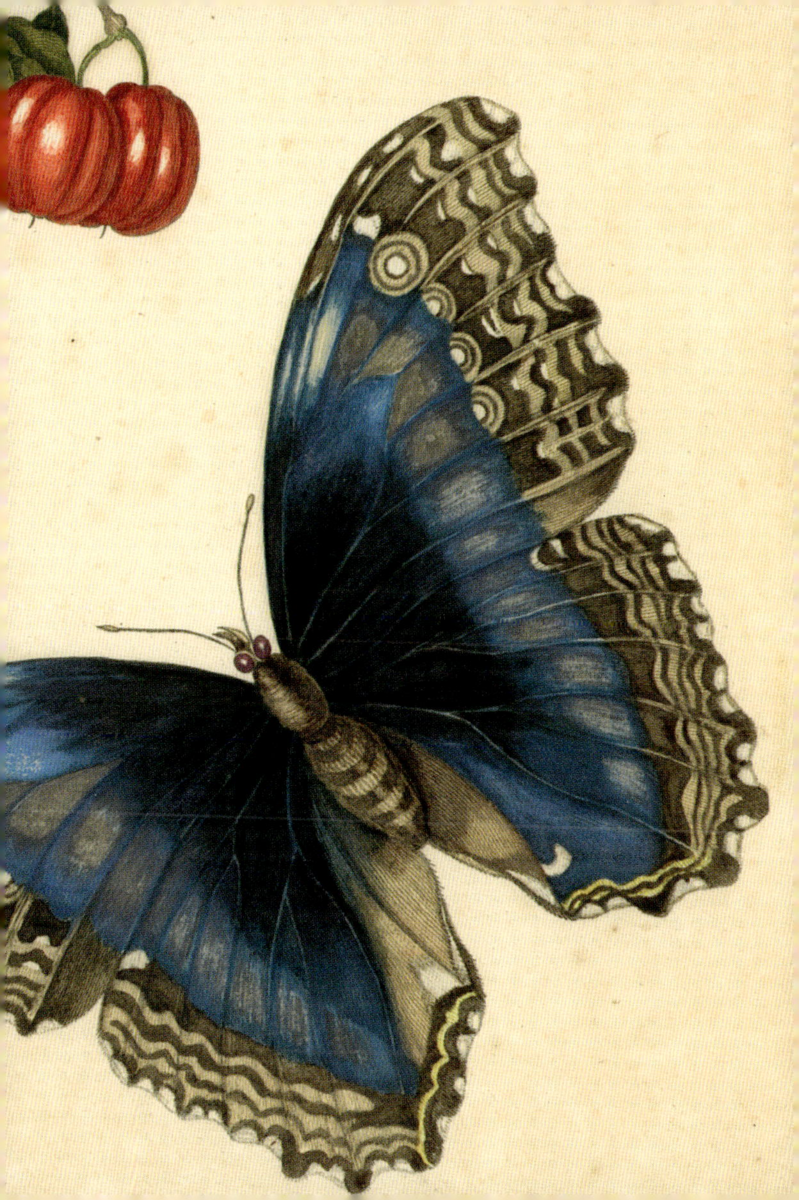

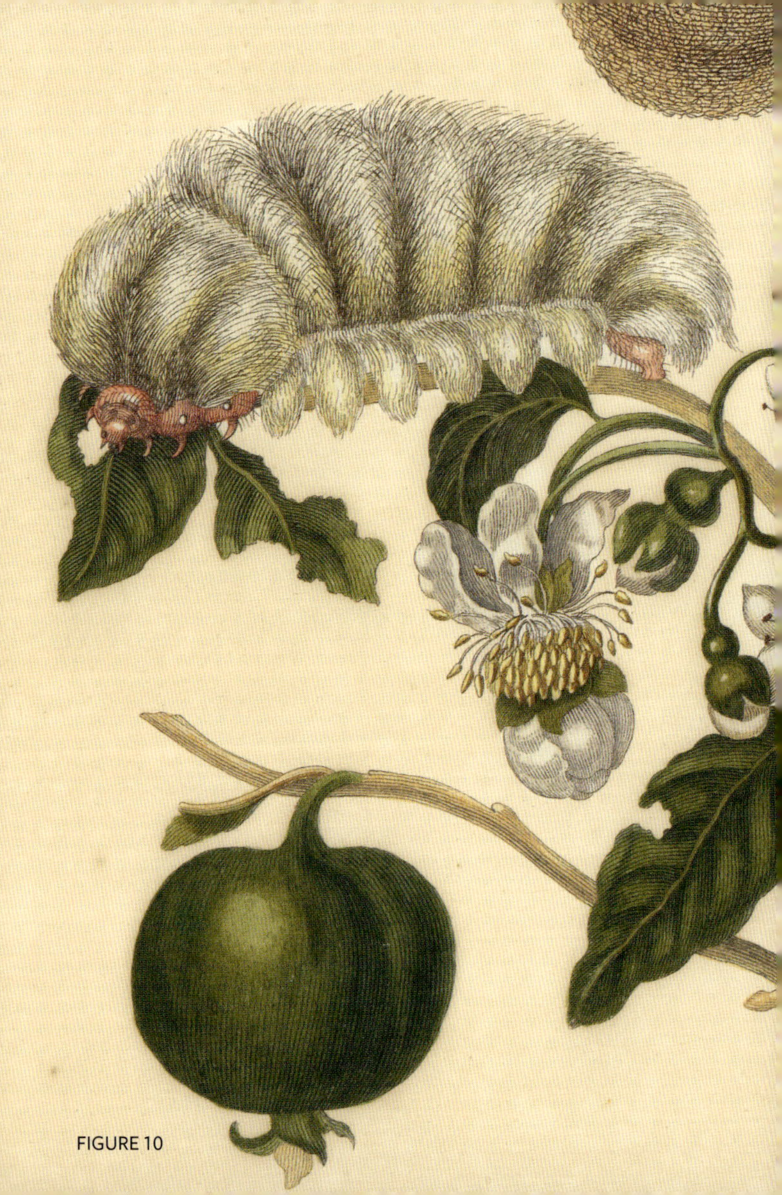

FIGURE 10

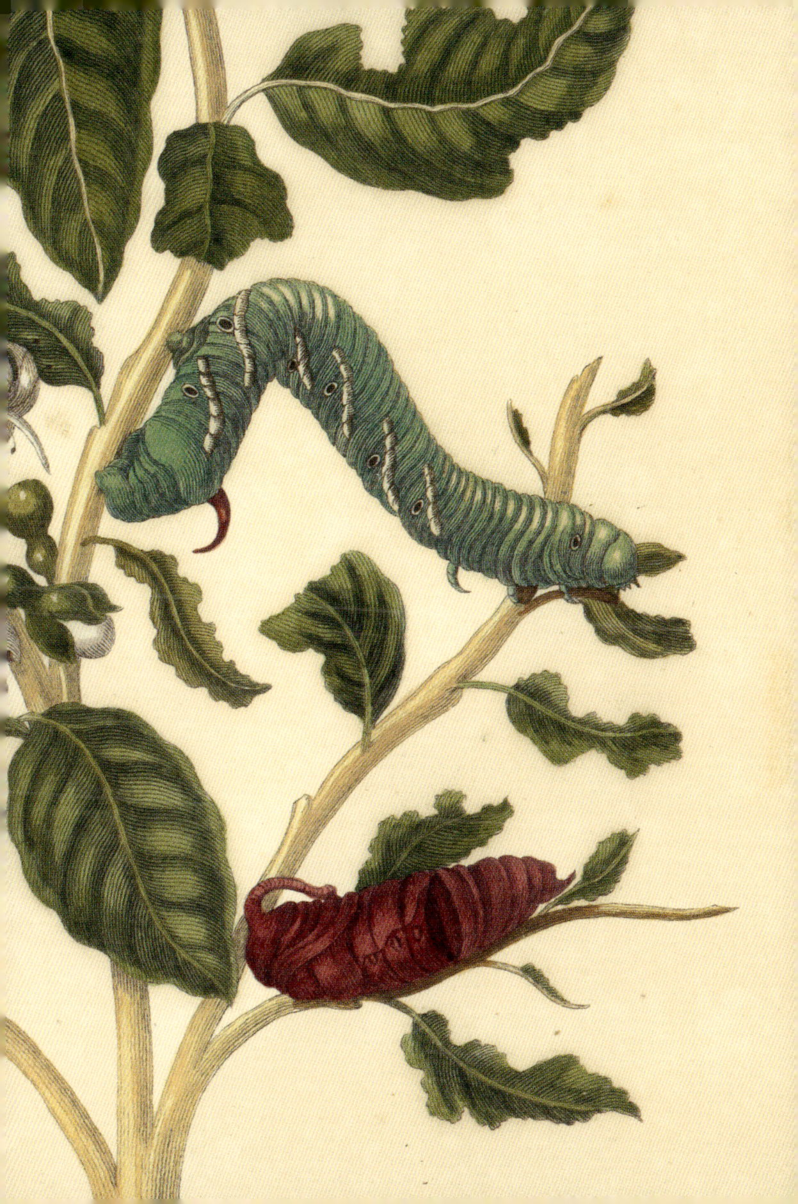

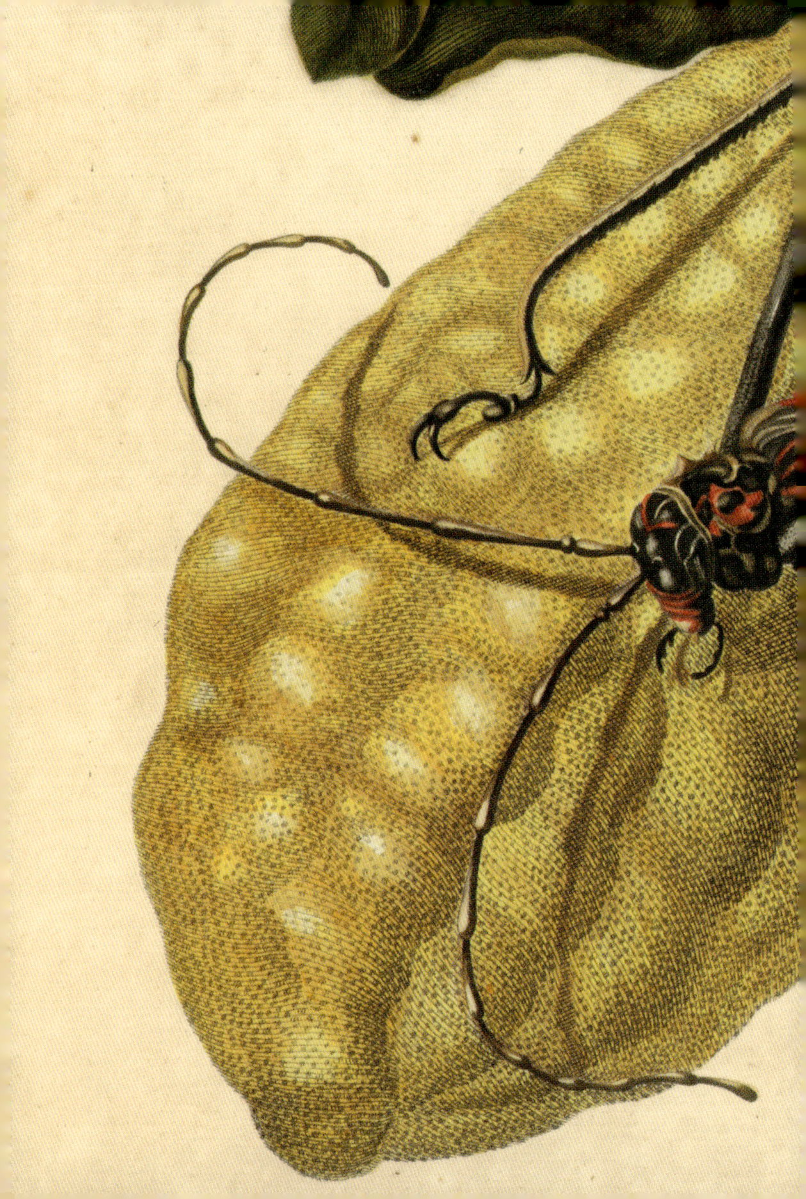

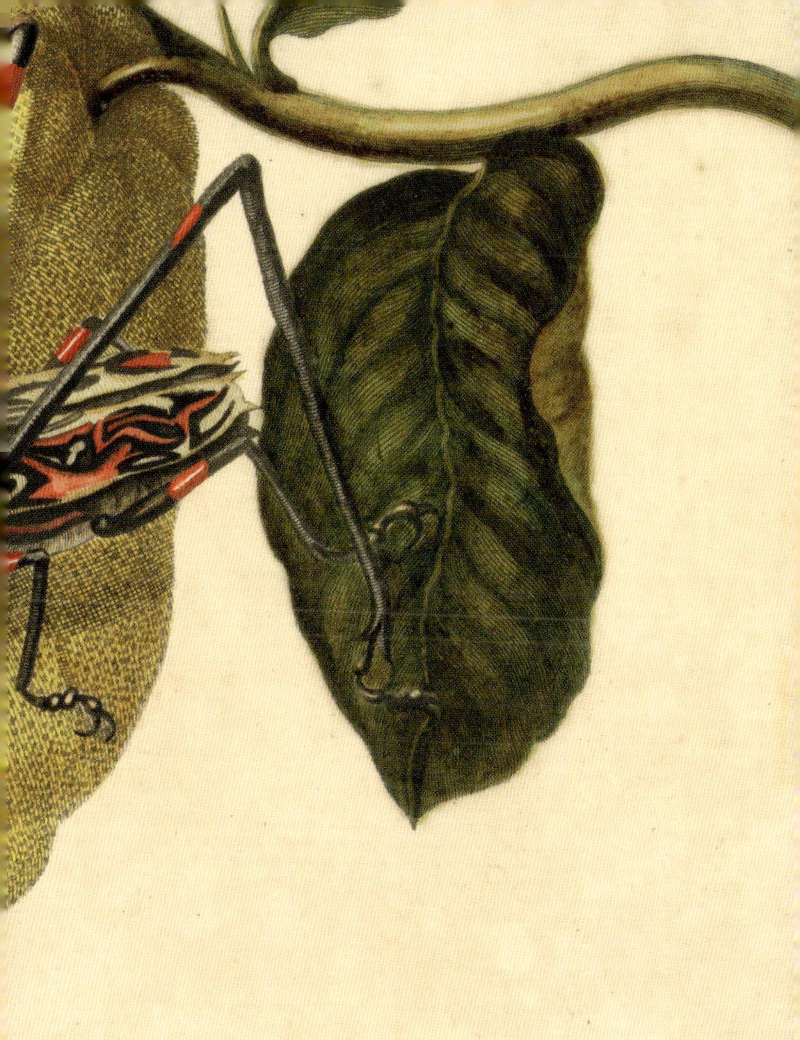

FIGURE 11

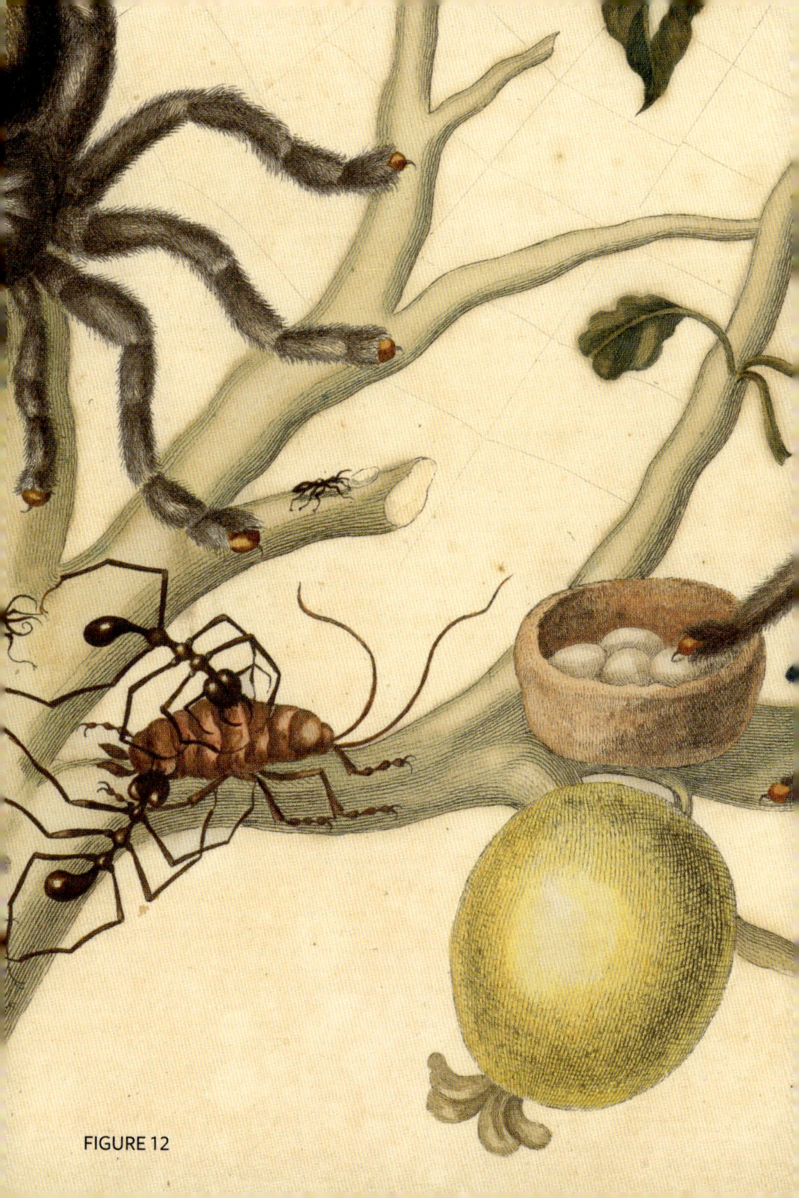

FIGURE 12

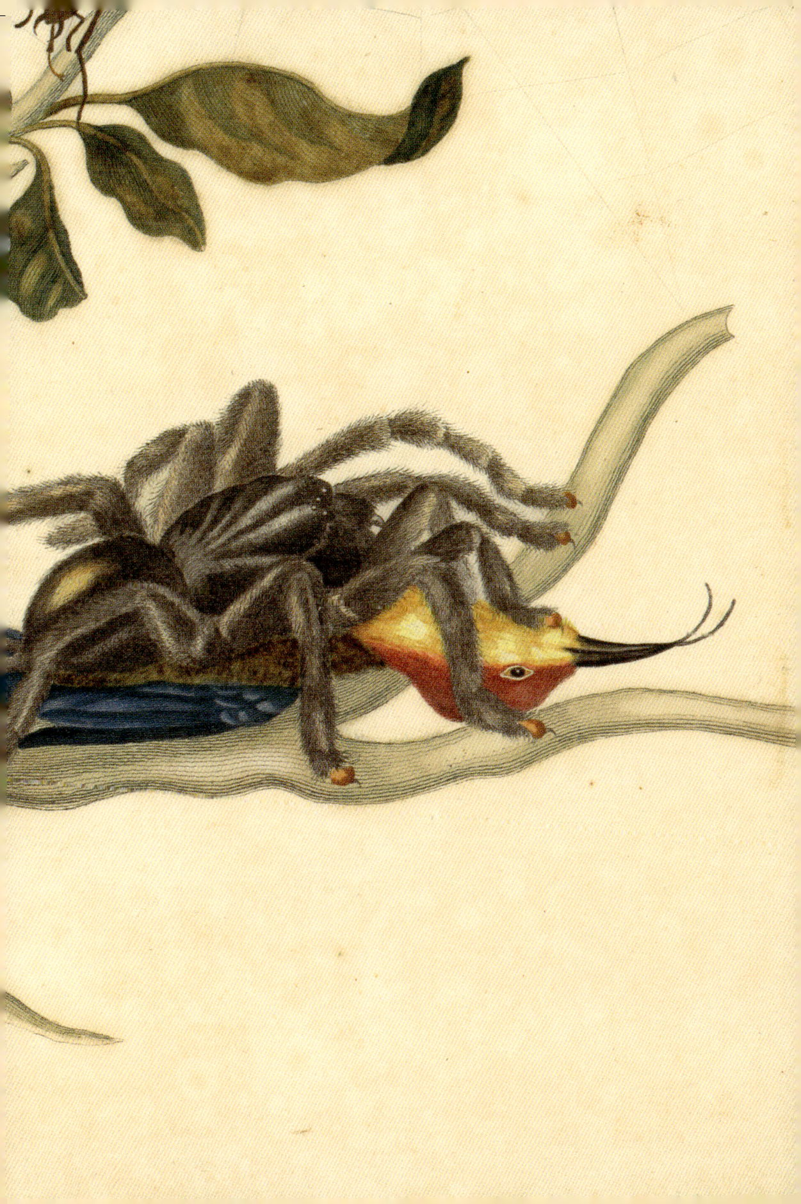

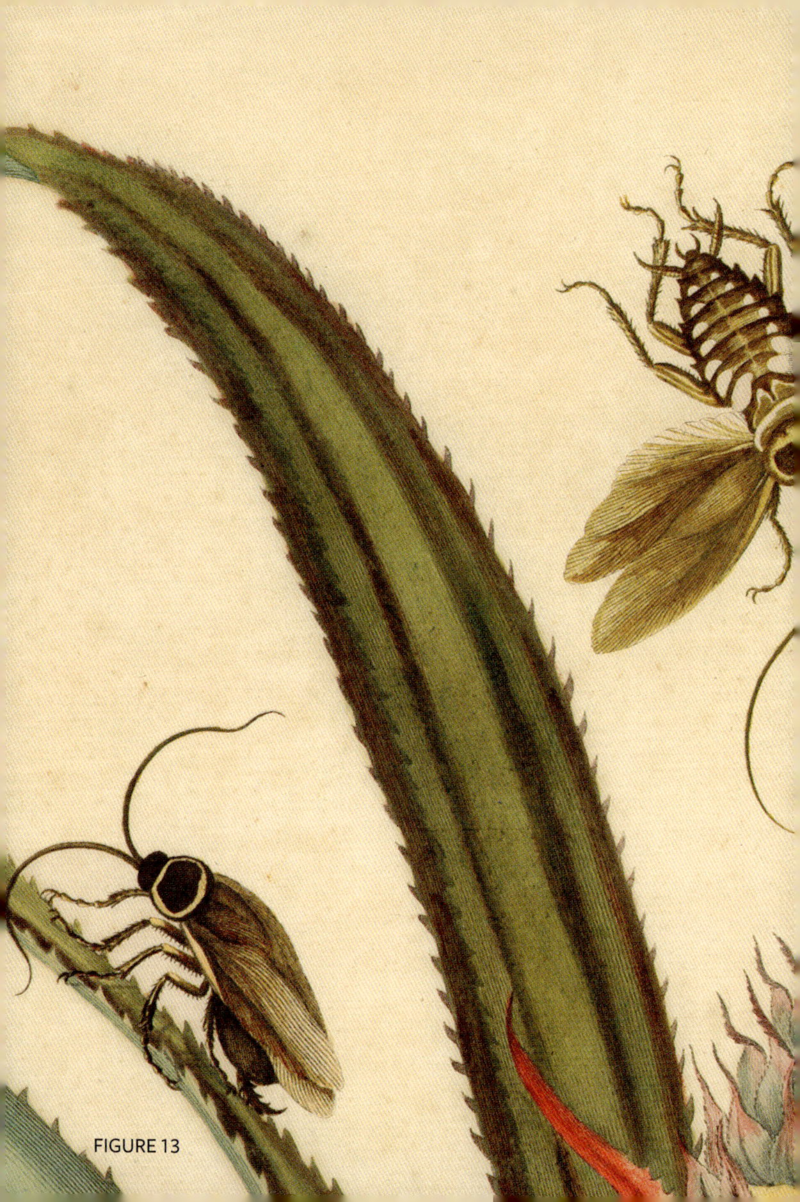

FIGURE 13

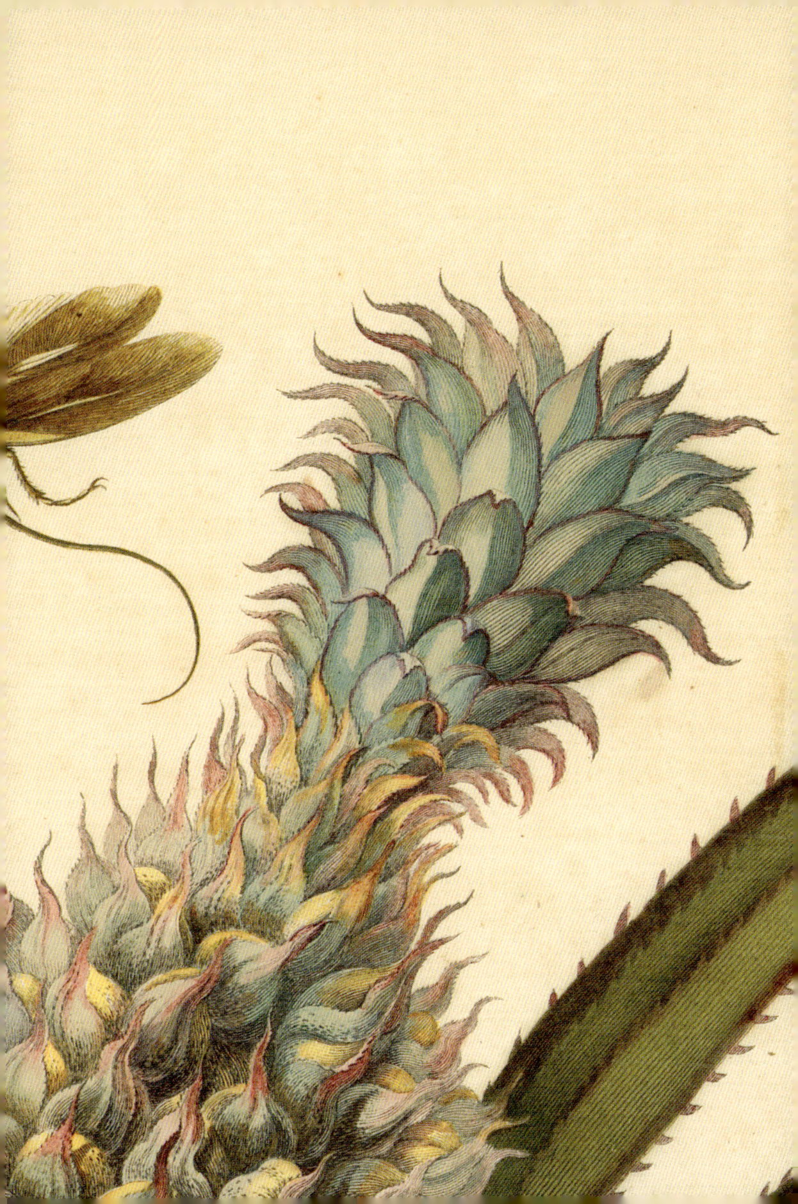

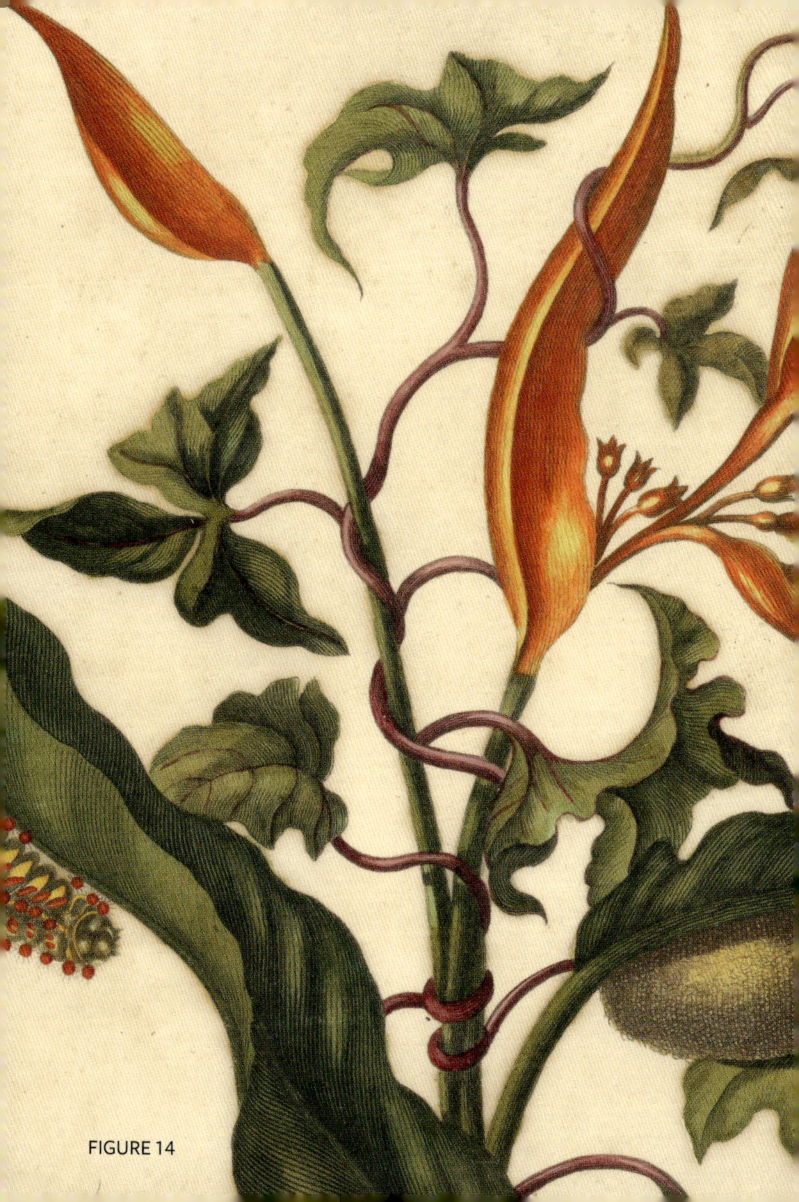

FIGURE 14

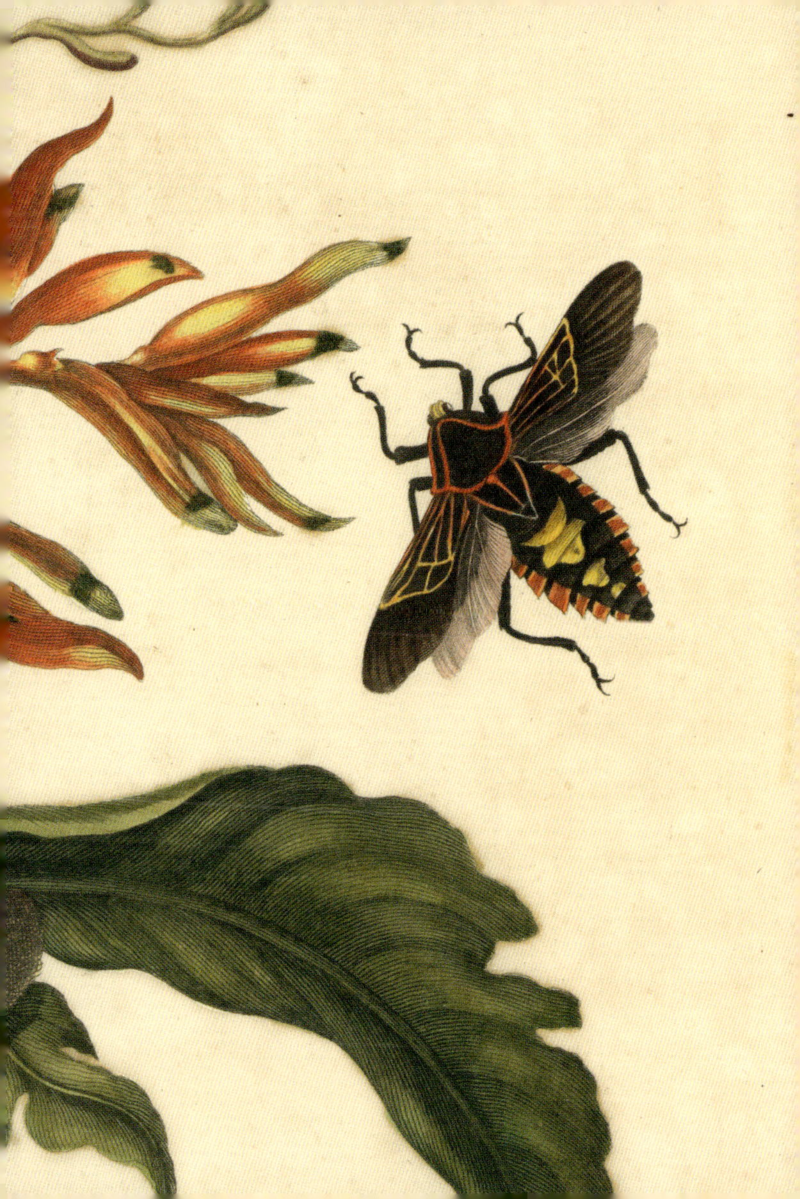

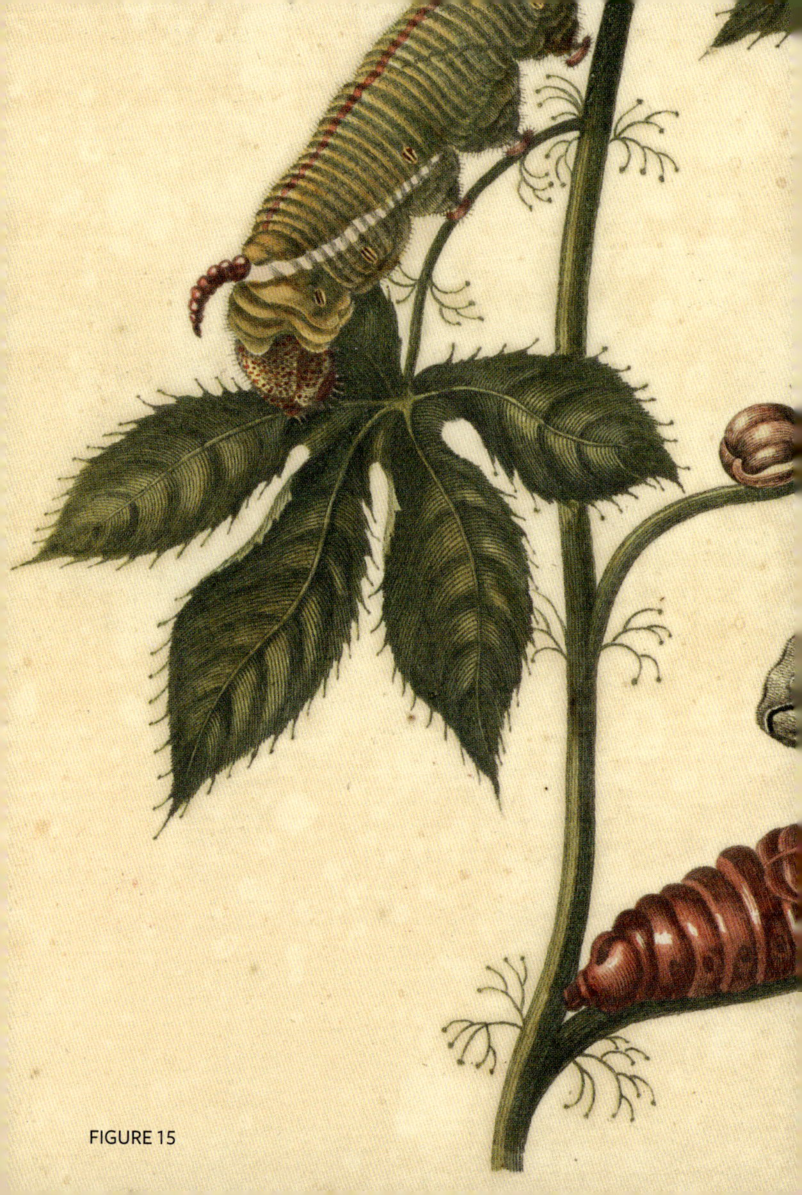

FIGURE 15

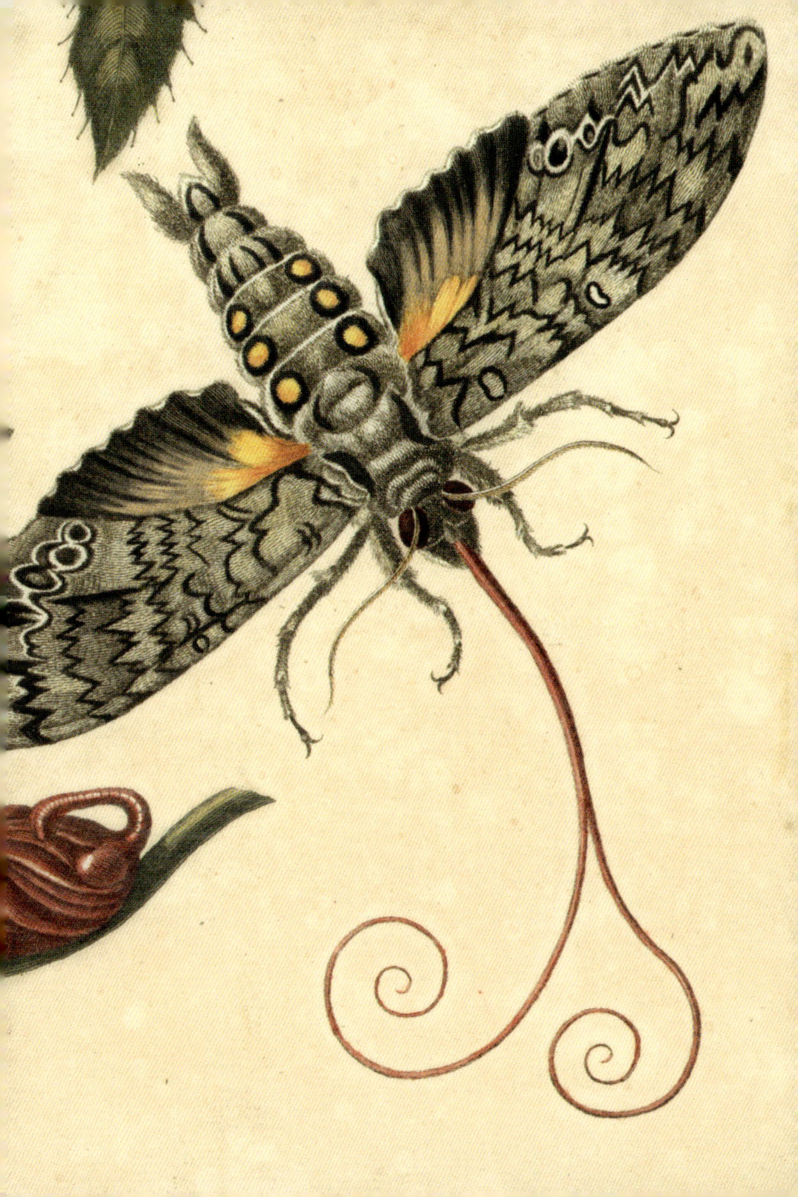

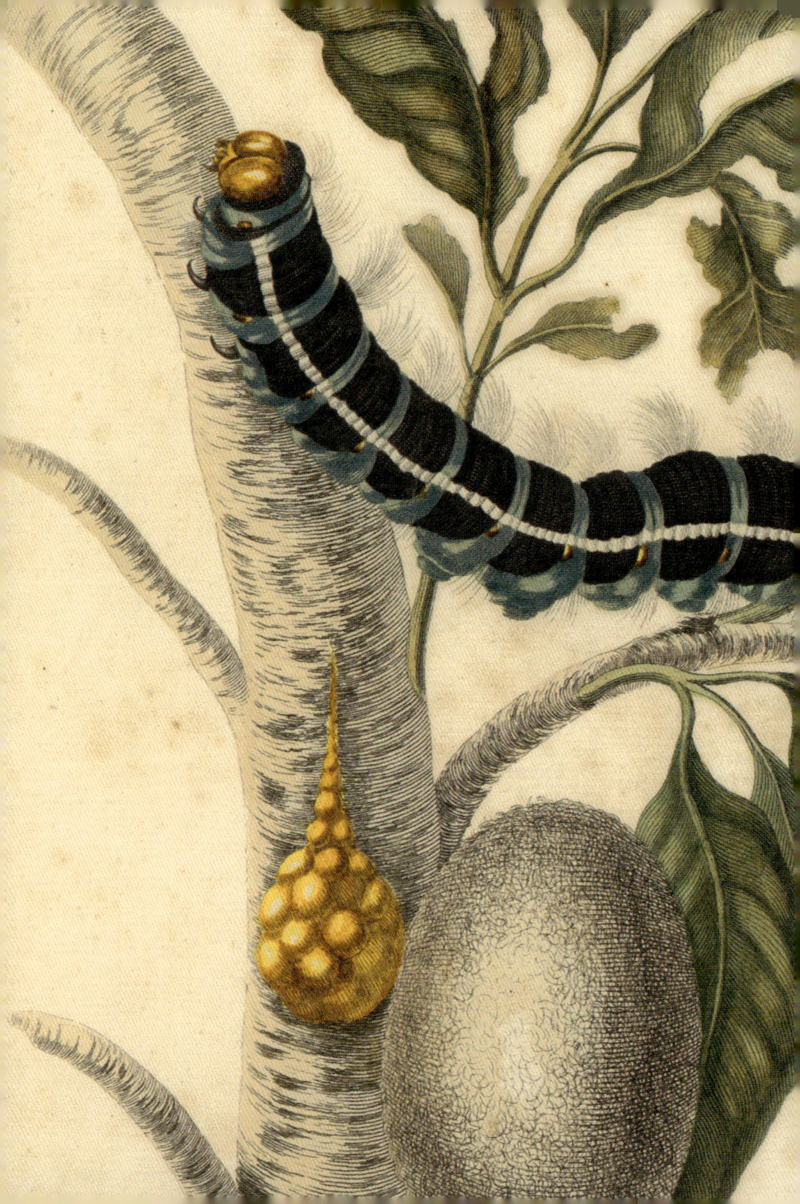

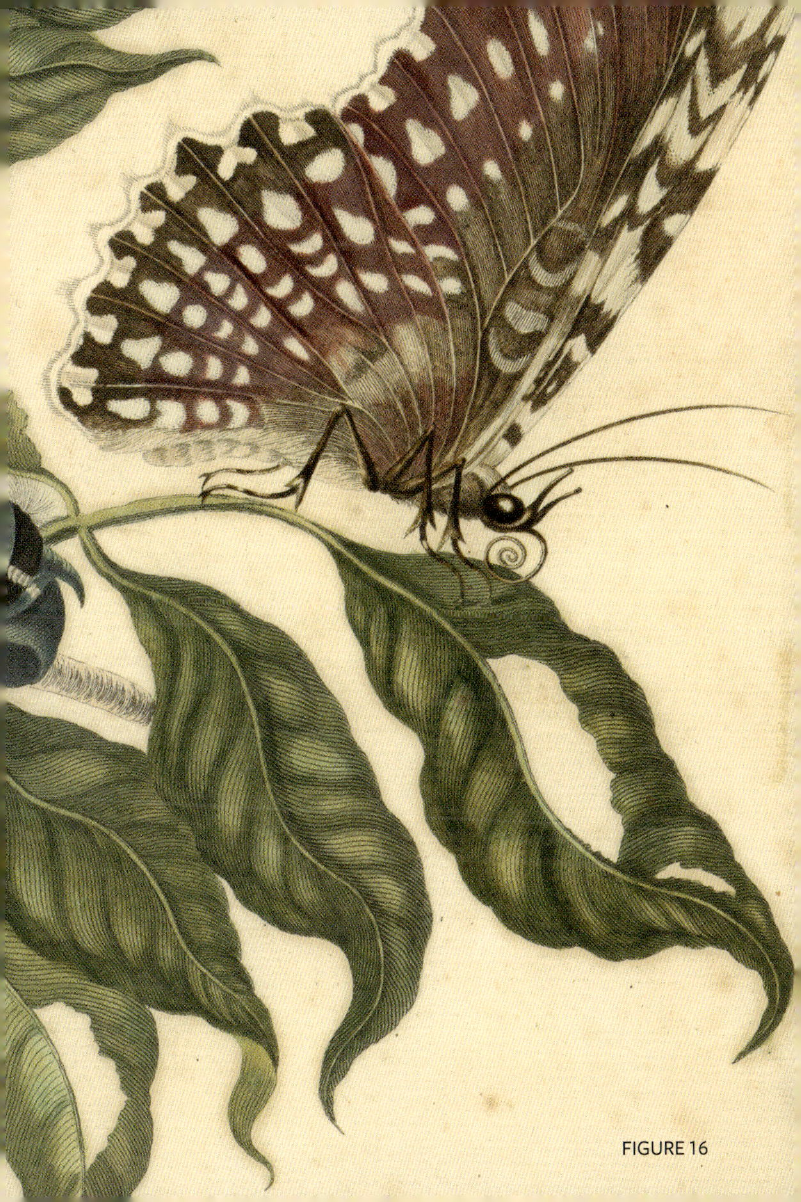

FIGURE 16

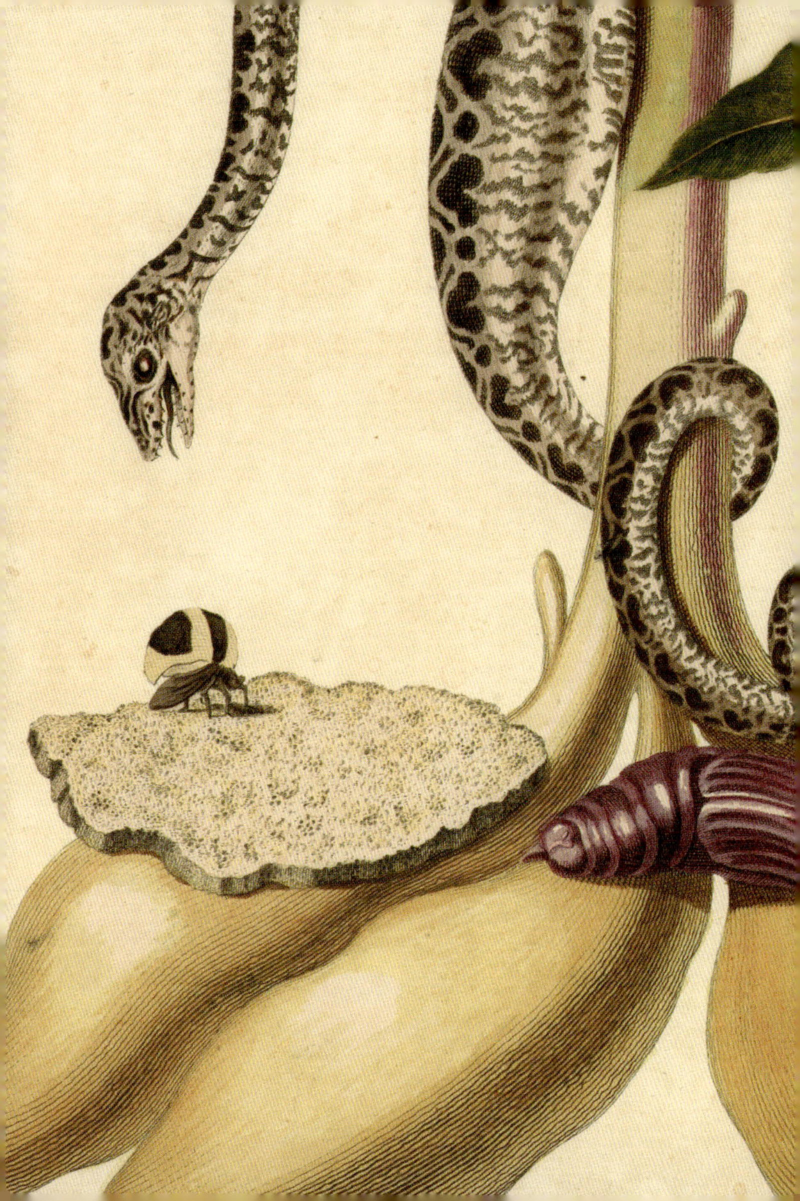

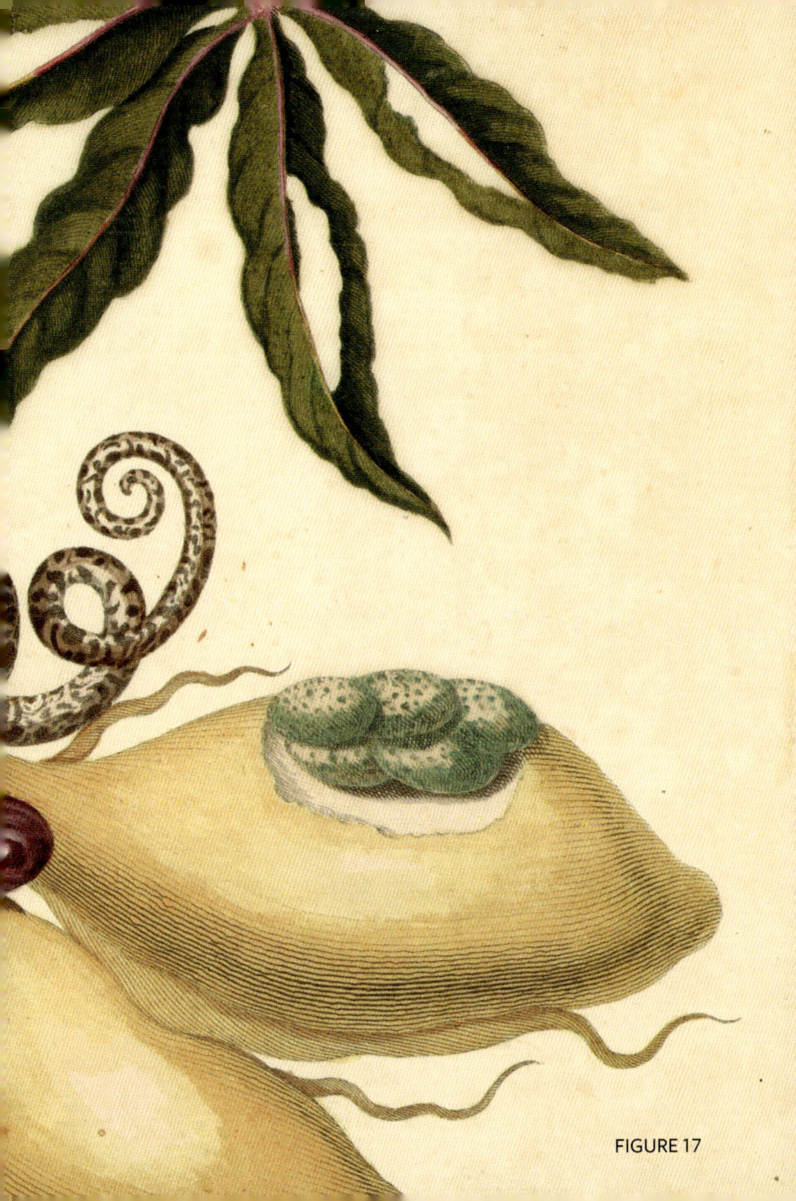

FIGURE 17

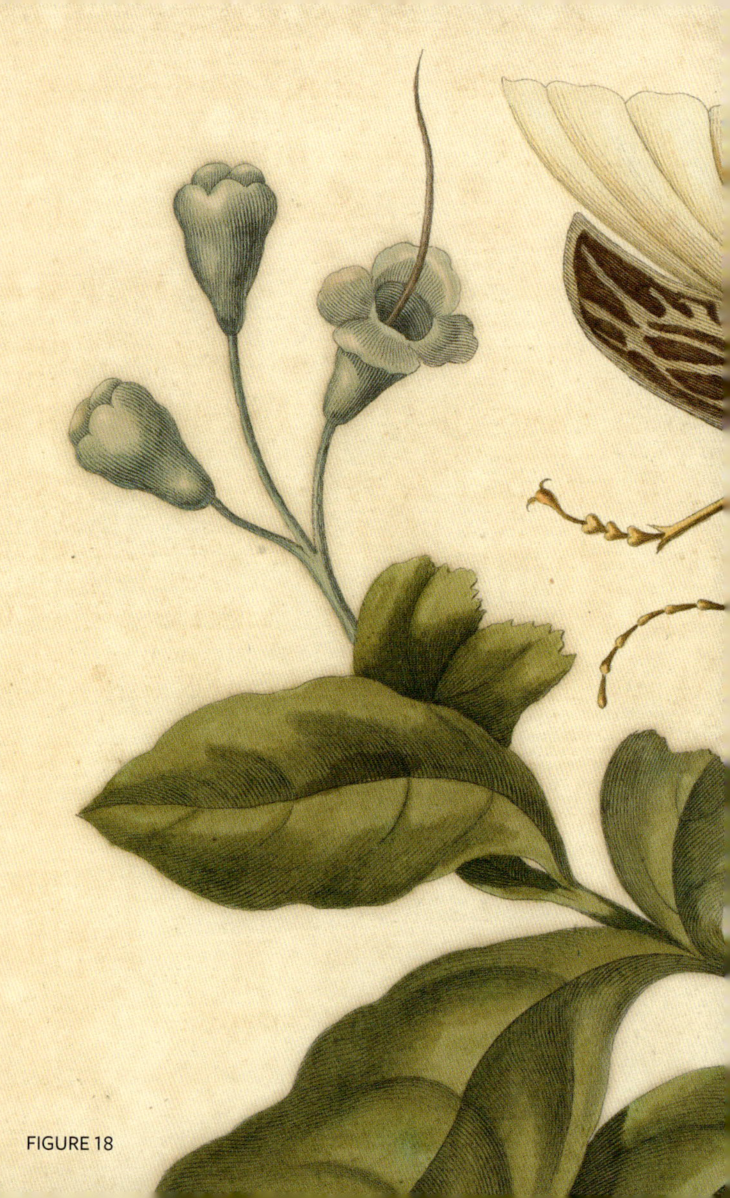

FIGURE 18

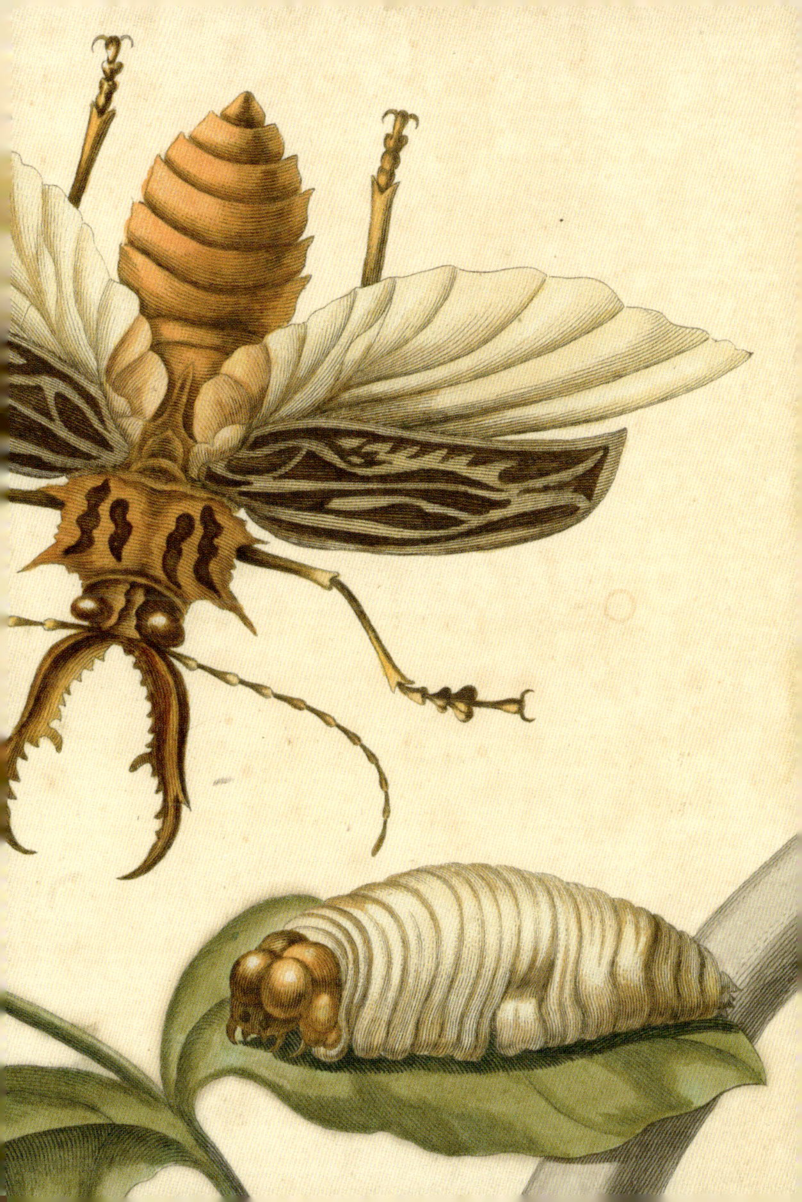

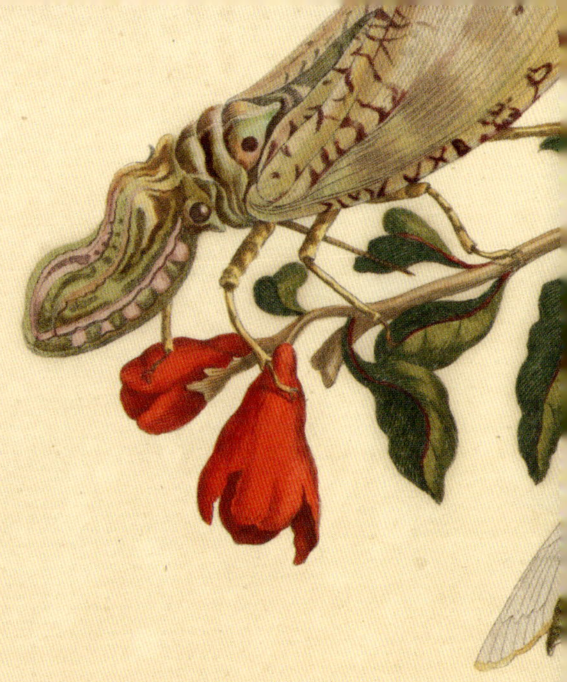

FIGURE 19

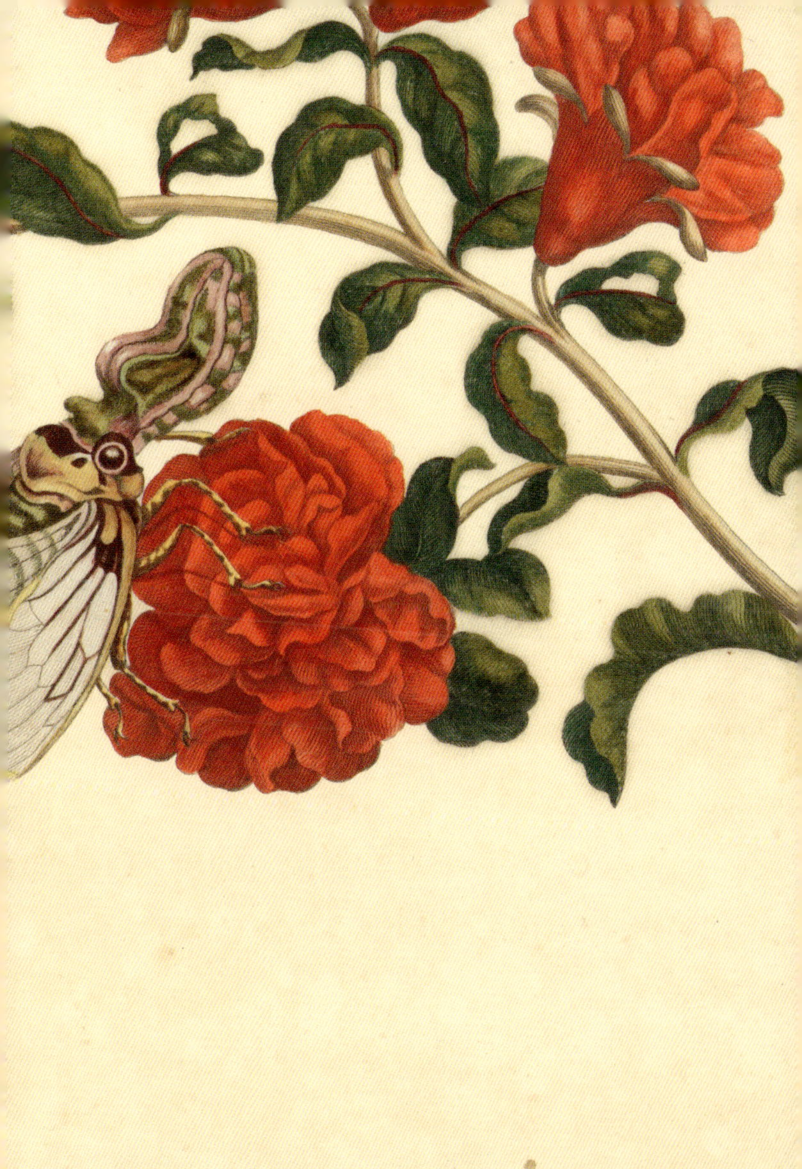

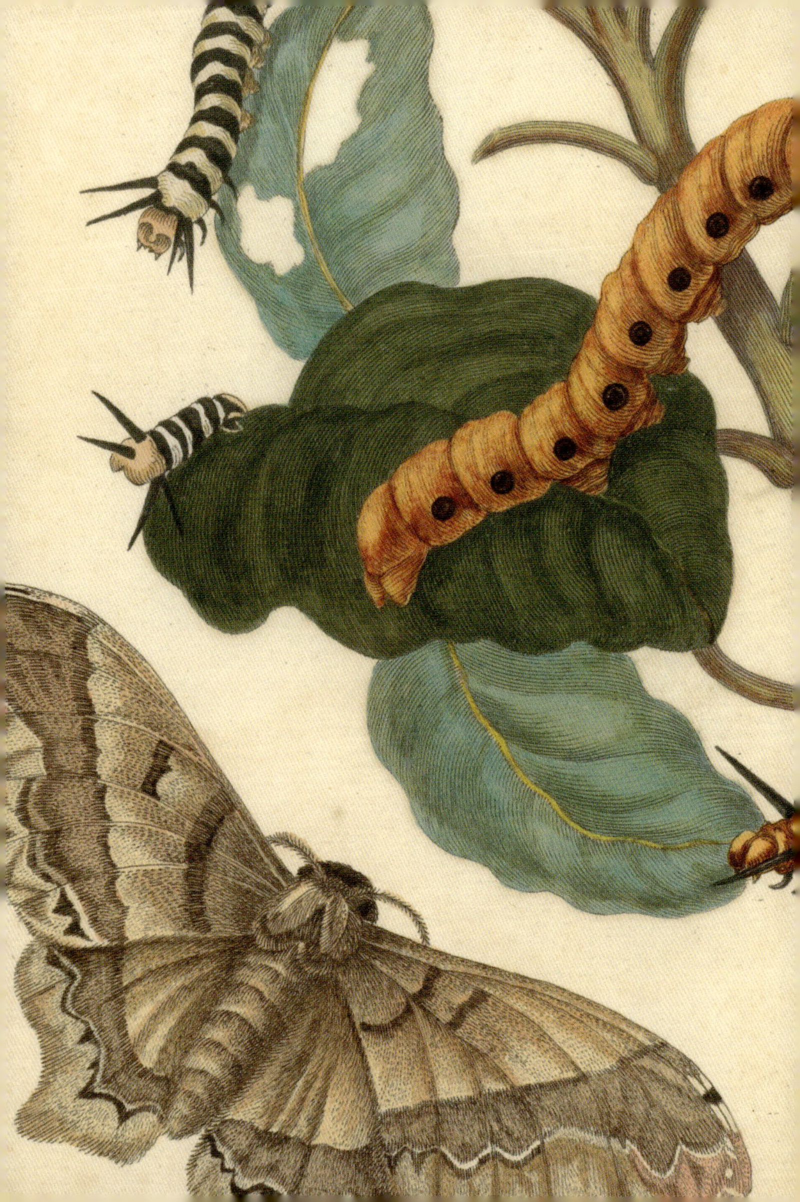

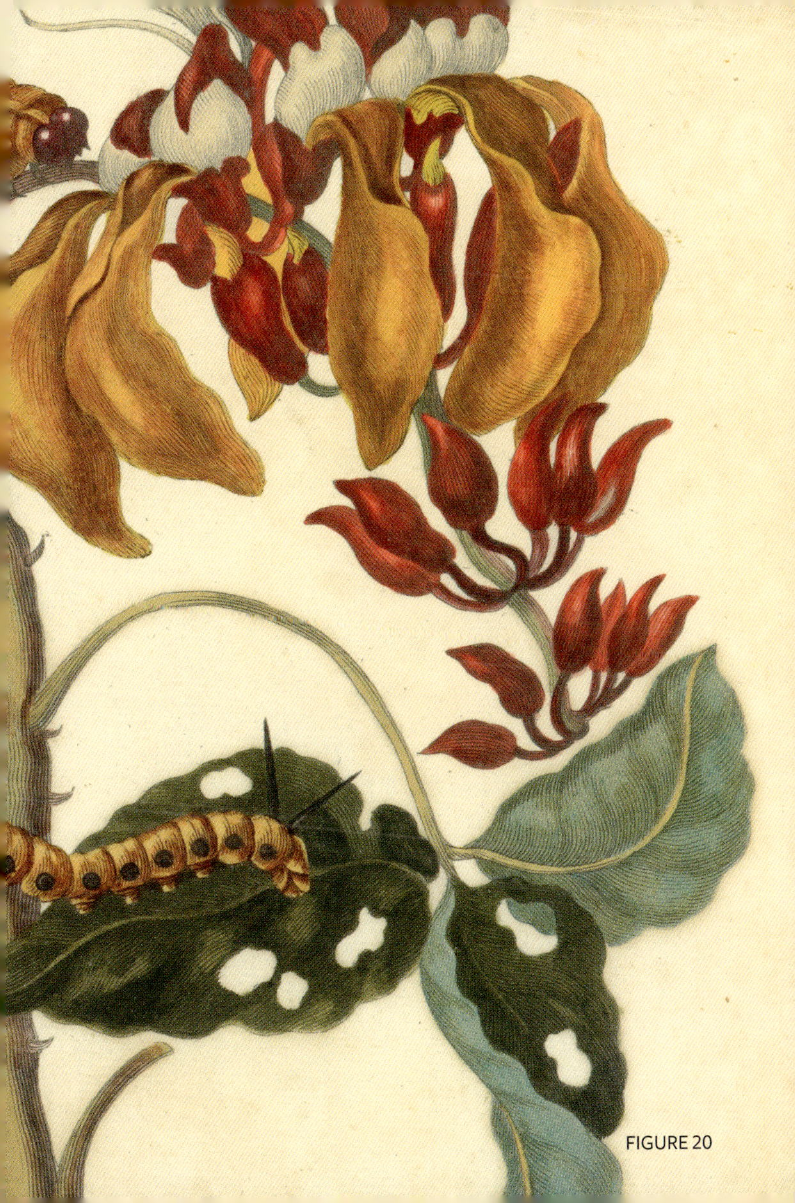

FIGURE 20

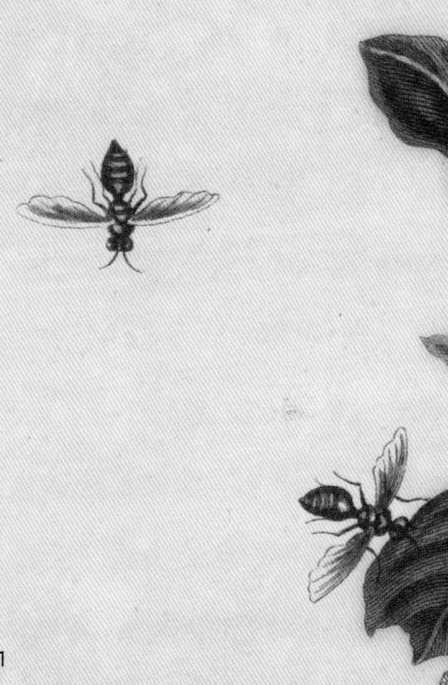

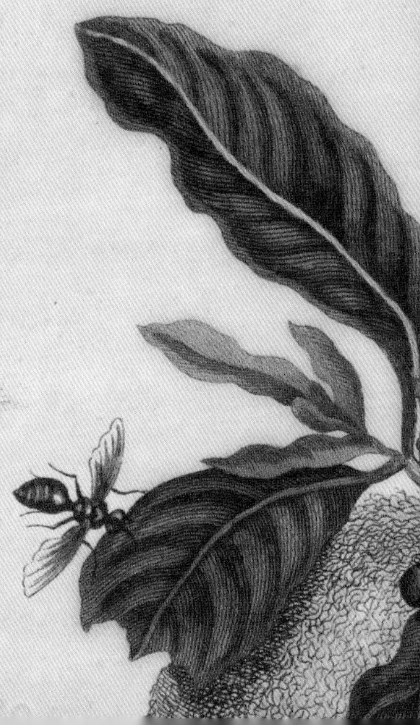

FIGURE 21

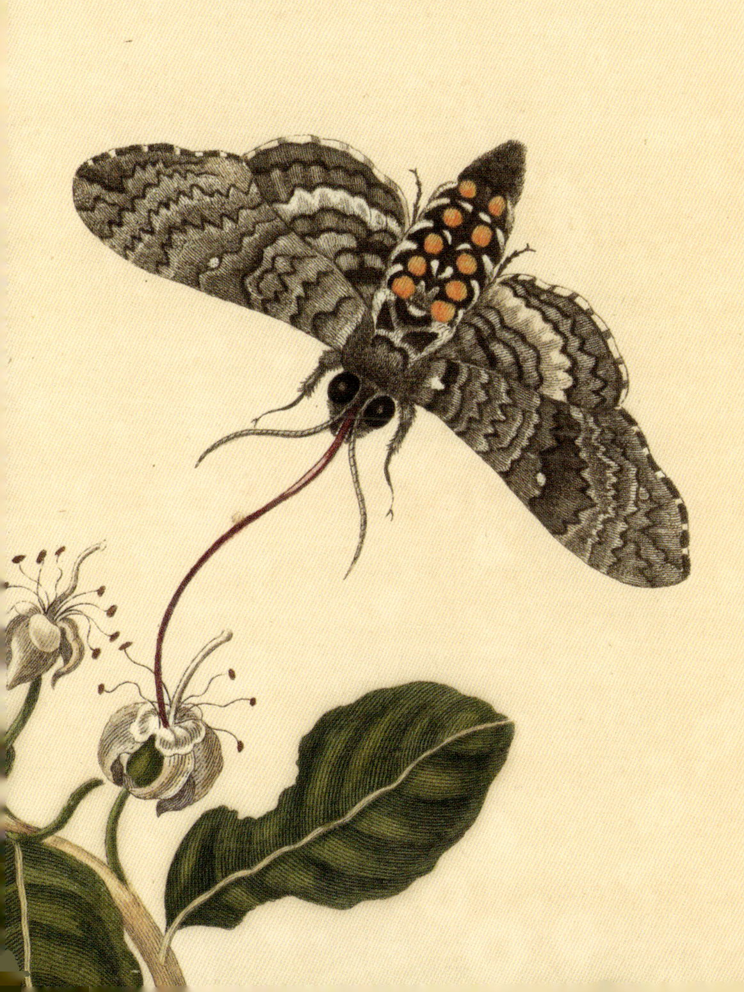

AFTERWORD

She started collecting caterpillars when she was thirteen. So began the career of Maria Sibylla Merian (German, 1647–1717), the woman who revolutionized the science of zoology and the art of illustrating it. Catching caterpillars in gardens, she made watercolors of their transformation into butterflies and kept a journal of her observations. When she was thirty-eight—by then a renowned artist-naturalist—she left her husband and joined a utopian religious community in the north of Holland. In 1699, aged fifty-two, divorced and living in Amsterdam, she sold most of her collection and took her younger daughter to the South American jungles of Suriname to capture in living color insect life and the fruit and flowers on which it fed.

According to Merian's introduction to her *Metamorphosis of the Insects of Suriname* (Amsterdam, 1705), the reasons for this critical change in her midlife were simple. Amsterdam in the 1690s was the hub of the global commercial empire of the Dutch East India Company, and Maria Sibylla's reputation had earned her easy access to the scholarly circles of naturalists and collectors, who acquired and studied the plants and insects brought back by traders from around the globe. The specimens she examined, however, were inanimate, devoid of any indication of the complex cycles of birth, growth, and decay that could be observed within their natural environment. "So I was goaded to undertake a huge and costly trip, traveling to Suriname in America, a hot and humid land where swarms of insects are there for the capture." And capture them she did, for posterity in watercolors and engravings.

Maria Sibylla Merian's obsession with exotic travel and the art of scientific illustration had been instilled in her from an early age. She was born in Frankfurt am Main in 1647, the daughter of Matthias Merian the Elder (German, 1593–1650), one of the most prolific engravers and publishers of his age, particularly admired for the scientific illustrations he executed for works on alchemy. His commercial success, however, came from the engravings he produced for the later parts of the *Great Voyages* (Frankfurt, 1590–1622), a series of lavishly illustrated volumes documenting the discovery and colonization of the exotic New World.

Matthias Merian died when Maria Sibylla was only two, and her mother married Jacob Marrel (German, 1614–81), an affluent still life painter, engraver, and art dealer. From her stepfather, Maria Sibylla learned how to draw, produce watercolors, and engrave on copper plates; she also grew up surrounded by the books, prints, and paintings left by her birth father.

In 1665 she married her stepfather's favorite pupil, Johann Andreas Graff (German, 1637–1701), with whom she had two daughters. She and her husband also collaborated on publishing her first work, *New Book of Flowers* (Nuremberg, 1675–80). In its preface, she expressed her hope that the pictures would strike a balance between art and nature. Maria Sibylla's creative desire echoed a sentiment found in captions for some of her father's engravings: "alchemy is the art that mirrors nature." For the Merians—both father and daughter—art was a form of spiritual devotion. Human creativity was an imitation of God's creative acts in shaping the scientific wonders of the world.

For her next publication, *Caterpillars, Their Wondrous Tranformation and Peculiar Nourishment from Flowers* (Nuremberg, 1679–83), she again collaborated with her publisher husband to fully explore her lifelong fascination with the transformation of caterpillars into butterflies. In 1685, two years after the appearance of this publishing milestone, Maria Sibylla left him. Apparently, an urge for a transformation in her own life spurred her to undertake a spiritual quest. She joined the Labadists, a radical Protestant commune founded on the precepts that commercial gain in the material world is the root of mortal sin and corruption. After six years, however, Maria Sibylla left the group and moved to Amsterdam, and then, after eight years, on to the Dutch colony of Suriname in June 1699 with her younger daughter, Dorothea, who was by then also a skilled artist.

The Merian women's journey was possibly the first purely scientific expedition to the Americas ever undertaken by individuals on a private initiative. Earlier scholars had been supported financially by nobility and accompanied expeditions of conquest and colonization. Here, mother and daughter traveled alone, and explored the South American rain forest for two years, studying and collecting, drawing and painting exotic plants and animals, recording all in Maria Sibylla's journal—the same one she started at thirteen.

The two women did not work entirely alone. They were assisted by African and Amerindian slaves from local Dutch plantations. Unlike most Dutch colonists, however, Maria Sibylla learned Carib, the language of the local Amerindians, and *Negerengels* (Black English), the Dutch name for the creole dialect developed by African slaves in the New World. Her attitude toward slavery can perhaps be detected in *The Insects of Suriname*. In her commentary to figure 3, Maria Sibylla explains that the berries depicted on a peacock flower are used by the native women to induce abortion and ensure that they will not bear offspring destined for a life of slavery. She adds that the slaves from Angola are treated harshly by the Dutch and dream of an afterlife "of peace and freedom" in their homeland.

When Merian and her daughter sailed back to Amsterdam in 1701, they were accompanied by an Amerindian woman who later collaborated in the preparation of their book. The first edition of *Metamorphosis of the Insects of Suriname* appeared in Amsterdam in 1705. Printed in a large folio format, it contained sixty full-page engraved plates, each accompanied by explanatory text on facing pages. Also available was a deluxe edition hand colored by the author and her daughters. For the hand-colored copies, Merian preferred the counter proof technique of pulling a print from the copperplate engraving. This method produced a softer line than the engraving, and minimally disrupted the hand coloring. The colored counterproof illustrations are closer and in the same direction as the watercolor originals on which they are based.

Plant and animal specimens were labeled with both native and Latin names (if they had been identified already in European scientific circles). These commentaries are also a treasure of historical anthropology. Merian documents the use and preparation of insects, frogs, plants, and flowers for food and medicine, from either first-hand observation or testimony of South American women who used these traditional methods to care for their families.

First and foremost, Maria Sibylla expressed herself pictorially. Figure 5 shows cockroaches around an unripened pineapple. Another pineapple is illustrated as ripe and bright yellow, populated by caterpillars, cocoons, and butterflies (figure 22). Thus, Merian introduces the lush sweetness of the exotic New World and evokes the perpetual cycle of life. A tarantula devours a female hummingbird alongside her nest while army ants decimate foliation (figure 12). A yellow caterpillar chewing on withering cherry leaves is rendered side by side with the blue and gold butterfly it will become (figure 9). Bud, flower, and fruit appear simultaneously along the stem of a guava tree (figure 10). Merian's priority is not to structure images in terms of scientific classifications of animal and plant species. Rather, she encapsulates a span of time into one artistic space, and depicts in a single image the full spectrum of a life cycle within its ecosystem. Indeed, she was among the first in history to take this approach to studying and illustrating the natural world.

According to the correspondence of Maria Sibylla, the 1705 first edition of *Metamorphosis of the Insects of Suriname* met with little commercial success. Plans to publish French and German translations for wider circulation throughout Europe had to be temporarily abandoned. Precious few copies of that first edition, colored by Maria Sibylla's hand, survive today in libraries around the world. Held in the Getty Research Institute is a copy of the second edition, which was published in Amsterdam in 1719 by Joannes Oosterwyk. It contains twelve additional plates, ten of which are based on original watercolors by Maria Sibylla not included in the first edition.

The second edition earned both commercial and critical success, and was reprinted in three subsequent editions. For the rest of the eighteenth century, Maria Sibylla Merian's work formed the indispensable foundation for all serious entomological research. Most notably, Carl Linnaeus (Swedish, 1701–78), regarded as the father of modern zoological classification, cited her hundreds of times in his research. In contrast, nineteenth-century British naturalists severely discredited her work. Recent scholarship, however, has proven that much of that criticism was unfounded and stemmed, in part, from the disbelief among nineteenth-century gentlemen-scientists that a seventeenth-century female artist could have produced reliable scientific data.

Consider in that context a final historical note: In 1669, the eminent biologist, Marcello Malpighi (Italian, 1628–94) published *De bombyce* (On the Silkworm) in which he announced to the European scientific community his discovery of the metamorphosis of silkworms into moths. This phemonemon had been noted already—nearly a decade earlier—in 1660. It was the first entry Maria Sibylla Merian recorded in her journal when she was thirteen.

DAVID BRAFMAN
Curator of Rare Books
Getty Research Institute

with the assistance of
STEPHANIE SCHRADER
Assistant Curator of Drawings
J. Paul Getty Museum

DETAILS OF PLATES

FROM MARIA SIBYLLA MERIAN'S *METAMORPHOSIS OF THE INSECTS OF SURINAME.*

Plates in parentheses correspond to Merian's original plate numbers.

COVER
Banana tree flower (*Musa paradisiaca*) with larva of io moth (*Automeris liberia*)
(Plate 12)

FIGURE 1
Banana tree (*Musa paradisiaca*) with metamorphosis of io moth (*Automeris liberia*)
(Plate 12)

FIGURE 2
Genip tree (*Genipa americana*) with South American palm weevil (*Rhynchophorus palmarum*), unidentified hairy larva (species of *Megalopygidae*), and bee (*Eulema cingulata*)
(Plate 48)

FIGURE 3
Peacock flower (*Caesalpinia pulcherrima*) and tobacco hawk moth and larva (*Manduca sexta*)
(Plate 45)

FIGURE 4
Thistle-like plant (*Solanum stramoniifolium*), moth (species of *Automeris*), and unidentified larva. (Plate 6)

FIGURE 5
Pineapple (*Ananas comosus*) with Australian cockroaches (*Periplaneta australasiae*) and German cockroaches (*Blatella germanica*)
(Plate 1)

FIGURE 6
Guava tree (*Psidium guineense*) with leafcutter ants (*Atta cephalotes*) and huntsman spiders (*Heteropoda venatoria*)
(Plate 18)

FIGURE 7
Cassava tree (*Manihot esculenta*) with rustic sphinx (*Manduca rustica*), larva of tetrio sphinx (*Pseudosphinx tetrio*), and garden tree boa (*Corallus enhydris*)
(Plate 5)

FIGURE 8
Double-blossomed pomegranate (*Punica granatum*) with lantern flies (*Fulgora laternaria*) and cicada (*Fidicina mannifera*)
(Plate 49)

FIGURE 9
Trinidad cherry tree (*Malpighia punicifolia*) with Achilles morpho (*Morpho achilles*) and unidentified larva
(Plate 7)

FIGURE 10
Guava (*Psidium guineense*) with hairy larva (species of *Podalia* or *Megalopyge*) and larva and pupa of tobacco hawk moth (*Manduca sexta*) (Plate 57)

FIGURE 11
Lemon (*Citrus medica*) and harlequin beetle (*Acrocinus longimanus*)
(Plate 28)

FIGURE 12
Guava tree (*Psidium guineense*) with army ants (species of *Eciton*), pink-toed tarantulas (*Avicularia avicularia*), and ruby topaz hummingbird (*Chrysolampis mosquitus*) (Plate 18)

FIGURE 13
Pineapple (*Ananas comosus*) with Australian cockroaches (*Periplaneta australasiae*) (Plate 1)

FIGURE 14
Parrot flower (*Heliconia psittacorum*) and sweet potato plant (*Ipomoea batatus*) with member of mesquite bug (*Pachylis pharanois*) and unidentified larva (species of *Eucleidae*) (Plate 41)

FIGURE 15
Known to Merian as the belly-ache bush (*Jatropha gossypifolia*) with metamorphosis of giant sphinx moth (*Cocytius antaeus*) (Plate 38)

FIGURE 16
Gumbo limbo (*Bursera simaruba*) and metamorphosis of giant Agrippa moth (*Thysania agrippina*) (Plate 20)

FIGURE 17
Cassava tree (*Manihot esculenta*) with pupa of tetrio sphinx moth (*Pseudosphinx tetrio*), garden tree boa (*Corallus enhydris*), and humpbacked cricket (*Membracis foliata*) (Plate 5)

FIGURE 18
Genip tree (*Genipa americana*) with longhorn beetle (*Macrodontia cervicornis*) and South American palm weevil larva (*Rhynchophorus palmarum*) (Plate 48)

FIGURE 19
Double-blossomed pomegranate tree (*Punica granatum*) with lantern fly (*Fulgora laternaria*) and fake insect with head of lantern fly attached to body of a cicada (Plate 49)

FIGURE 20
Swamp immortelle (*Erythrina fusca*) with saturniid moth (*Arsenura armida*) and larvae of unidentified species (Plate 11)

FIGURE 21
Guava (*Psidium guineense*) with tobacco hawk moth (*Manduca sexta*) and unidentified fly (Plate 57)

FIGURE 22
Pineapple (*Ananas comosus*) with metamorphosis of bamboo page (*Philaethria dido*) and twice-stabbed ladybird beetle (*Chilocorus cacti*) (Plate 2)

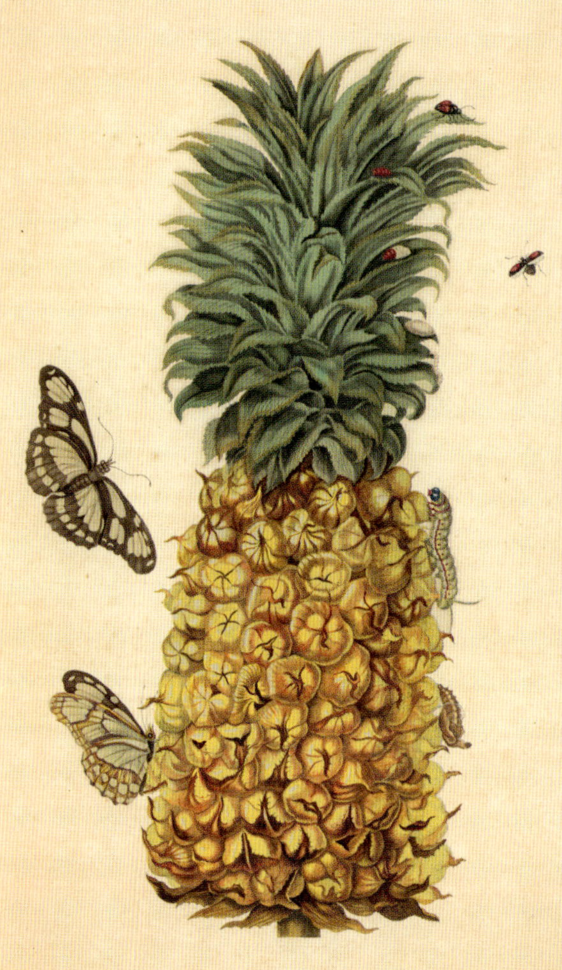

FIGURE 22